CONTEMPORARY TECHNOLOGIES FOR SHALE-GAS WATER AND ENVIRONMENTAL MANAGEMENT

T0308719

Contemporary Technologies for Shale-Gas Water and Environmental Management

About WEF
Formed in 1928, the Water Environment Federation® (WEF®) is a not-for-profit technical and educational organization with members from varied disciplines who work towards WEF's vision to preserve and enhance the global water environment.

For information on membership, publications, and conferences, contact

Water Environment Federation
601 Wythe Street
Alexandria, VA 22314-1994 USA
(703) 684-2400
http://www.wef.org

Contents

ii

FOREWORD

This document evolved from a set of back-to-back invited sessions focusing on shale-gas environmental technologies and related considerations. This document and associated sessions were prepared under the auspices of the Industrial Wastewater Committee of the Water Environment Federation. The committee provides technical information exchange, professional networking opportunities, communication of regulations, and policy and guidance. Programs addressing industrial and hazardous wastes treatment, disposal, prevention, remediation, reuse, and minimization are developed by the committee. For WEFTEC 2012, the committee sponsored a series of presentations on industrial issues, including a symposium relating to managing water and wastewater from shale gas extraction.

The symposium organizing committee included

Charles Meyer – Shell Corporation
Joe Cleary – HDR Qualcom
Steve Gluck – Dow Chemical
Ronald D. Neufeld – University of Pittsburgh

The committee was assisted by interested others who helped in the solicitation of papers. These included:

Aaron Frantz – CDM
Ajit Ghorpade – Veolia NA
James DeWolfe – Arcadis Corp
Tom Pederson – CDM

The gas-shale industry has the potential to bring significant economic gains and decrease U.S. dependence on foreign energy. On the other hand, early developments in the Marcellus Shale play have been associated with severe public health and environmental health consequences as well as heightened public sensitivities.

The industry is consolidating, with large energy companies purchasing drilling rights from numerous smaller and often undercapitalized entities. Research, technology developments, and renewed regulatory activities are currently underway, with the goal of correcting errors of the past and ensuring an environmentally acceptable shale-gas industry.

The purpose of these WEFTEC sessions is to bring together cutting edge gas-shale corporate developers, environmental researchers, and concerned municipality representatives to present contemporary technical information and explore potential issues and concerns with gas-shale stakeholders. The sessions are focused on related science and engineering technology and present information on emerging technologies, case studies, and rational cautions for chemical and radiological adverse human and environmental interactions.

Session specifics include

- Overview of gas-shale drilling and technology *(Fracking 101)* and management of water needs, reuse, and disposal.
- Overview of chemical aspects and significant considerations of shale-gas water management.
- Fate and transport of radium brought to the surface during drilling and other associated shale-gas operations.
- Relationship of shale formation properties, produced water quality, and water management strategies.
- The management of divalent metal scale formers *(water softening)* to enhance reuse of flowback water.
- Application of lessons learned from the engineered management of oilfield produced waters to shale-gas waters.
- In-process reuse and recycle of impaired shale-gas waters prior to disposal.
- Laboratory and field data linking conversion of bromides to bromate in conventional drinking water plant chlorination facilities.
- Reversible sorption and desorption of organics from impoundment basins and wells by an innovative organic–inorganic material based on silica.
- Conventional pretreatment and advanced polishing technologies available for produced water treatment and reuse.
- Case studies using nonconventional technology for treatment, reuse, and disposal of shale gas waters in Pennsylvania.
- Impoundment liner considerations for flowback water retention.

It is clear that rapid progress, as exemplified by the contributions to this symposium, is leading to a better understanding, development, and implementation of environmental technologies associated with shale-gas extraction and transport. This progress should do much to focus and alleviate potential public and regulatory concerns that can affect the future of the shale-gas industry.

Ronald D. Neufeld, Ph.D., P.E., BCEE, WEF Fellow
University of Pittsburgh
Symposium Coordinator and Proceedings Editor

OVERVIEW OF SHALE GAS WATER ISSUES

John A. Veil
Veil Environmental, LLC, Annapolis, Maryland, USA
john@veilenvironmental.com

ABSTRACT

Tens of thousands of wells are being drilled each year in several large gas shale formations in the United States. Shale gas development is also beginning in other countries and will likely increase rapidly. All U.S. shale gas wells must be fractured to allow sufficient gas to be produced to make the wells economically viable. In order to conduct fracturing operations, gas companies must obtain several million gallons (5,000 to 20,000 m^3) of water for each well from local sources. In some areas, the available freshwater supplies are limited, creating a challenge for the gas companies. After the frac job is finished, a portion of the frac fluid that was injected returns to the surface over the next few days. This "flowback water" is considerably saltier than the original frac fluid, plus it often contains elevated concentrations of metals, radionuclides, and other contaminants. This paper describes the water needs for making up the frac fluids for each well and how the necessary volume compares with other existing water uses in those regions. The steps in the shale gas process in which water is used and wastewater is generated are discussed, with an emphasis on management of the flowback water and smaller volumes of ongoing produced water from the shale formation. Various chemical additives are mixed with water and proppants to make frac fluids. The paper describes the general categories of chemicals that are used and provides information on the Frac Focus chemical registry developed to provide public access to the names and quantities of chemicals actually used in the frac fluids for many U.S. shale gas wells.

KEYWORDS: shale gas, hydraulic fracturing, flowback water, produced water, Frac Focus

INTRODUCTION

Production of natural gas from shale formations has become among the fastest growing sectors of the oil and gas industry. Estimates published by the U.S. Department of Energy's (DOE's) Energy Information Administration in January 2012 (EIA 2012) project that shale gas will steadily expand over the next 23 years and by 2035 will make up 49% of the U.S. gas production (see Figure 1). This projection is higher than last year's projection of 45%.

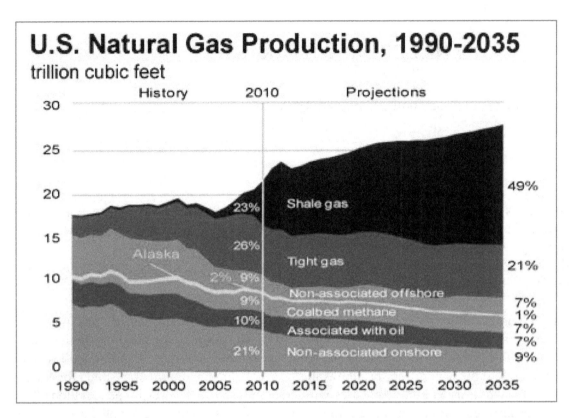

Figure 1. Natural Gas Production by Source, 1990-2035 (trillion cubic feet). Source: EIA (2012). Note that 1 trillion cubic feet = 28.3 billion cubic meters.

Although the industry recognized that many shale formations held substantial quantities of natural gas, conventional vertical wells were unable to produce significant quantities of gas. Shale gas production expanded in the Barnett Shale in Texas during the late 1990s after producers began employing two technologies on each well: a) horizontal drilling to allow longer contact between the well bore and the shale rock; and b) hydraulic fracturing to develop a network of cracks in the shale rock that allowed the gas to move to the well.

After shale gas production was shown to be feasible in the Barnett Shale, producers moved to other U.S. formations including:

- the Fayetteville Shale in Arkansas,

- the Haynesville Shale in Louisiana,

- the Woodford Shale in Oklahoma,

- the Marcellus Shale in Pennsylvania, New York, and West Virginia, Ohio, and Maryland,

- the Bakken Shale in North Dakota,

- the Eagle Ford Shale in south Texas, and

- the Utica Shale, which underlies much of the areal extent of the Marcellus but at greater depth.

2

A report commissioned by EIA and prepared by Advanced Resources International in spring 2011 surveyed other shale formations around the world to estimate their economic viability for natural gas production (ARI 2011). The report assessed 48 shale gas basins in 32 countries, containing almost 70 shale gas formations. These assessments cover the most prospective shale gas resources in a group of countries that demonstrate some level of relatively near-term promise and for basins that have a sufficient amount of geologic data for resource analysis. The report shows that the international shale gas resource base is vast. The initial estimate of technically recoverable shale gas resources in the 32 countries examined is 163 trillion m^3 – more than six times higher than the current U.S. technically recoverable shale gas estimate. It is highly likely that shale gas production will expand to other countries over the next decade.

STEPS IN THE SHALE GAS PROCESS

Figure 2 lists eight steps that make up the process for developing and producing shale gas. The shaded boxes represent steps that involve water in some manner. Each of these is discussed in the following sections.

| Gaining Access to the Gas (Leasing) |
| Searching for Natural Gas |
| Preparing a Site |
| Drilling the Well |
| Preparing a Well for Production (Hydraulic Fracturing) |
| Gas Production and Water Management |
| Moving Natural Gas to Market |
| Well Closure and Reclamation |

Figure 2. Steps in Shale Gas Production.

Preparing a Site

Before well drilling can begin, the company must clear vegetation, level enough land for the well pad, and construct a pad for the drilling rig and other equipment used in preparing the well. Drill pad sites are often located some distance from public roadways. Therefore, operators often must also construct an access road between the public road and the well site. Unless pitless drilling techniques are used, operators construct lined pits to hold drilling fluids and drill cuttings.

All of these excavation and grading activities expose soil to rainfall runoff. The operators take measures to minimize runoff by stabilizing the surface with gravel or other material. At some sites the center section of the pad, where the heavy equipment and wellhead are located, is covered with a geotextile liner material. Often ditches are constructed around the perimeter of the disturbed area to capture clean stormwater and move it around the site without becoming contaminated.

Drilling the Well

Shale gas wells are constructed with multiple layers of pipe known as casing. Shale gas wells are not drilled from top to bottom at the same diameter but rather in a series of progressively smaller-diameter intervals. First, a relatively large diameter hole is drilled from the surface down to about a 15-meter depth. This is lined with large casing called conductor pipe; the conductor pipe is cemented in place. Next, a smaller-diameter hole is drilled to a lower depth inside the previous string of casing. Another casing string is installed to that depth and cemented. This process may be repeated several more times. The final number of casing strings depends on the regulatory requirements in place at that location. It reflects the total depth of the well and the strength and sensitivity of the formations through which the well passes.

Nearly all shale gas wells use horizontal drilling. The well is drilled vertically until a few hundred meters above the shale formation. At that point, a drilling motor is placed on the leading end of the rotating drill string. The drilling motor gradually bends the well bore until the desired angle is reached. The horizontal section of the well can extend out more than 1,000 meters laterally from the vertical section of the well.

Water is required to make up the liquid component of the drilling fluid or mud that is used to lubricate the rotating drill bit. Drilling mud is pumped downward through the hollow drill pipe and exits through holes in the bit. The mud helps to convey the ground-up rock (drill cuttings) to the surface through the annular space between the drill pipe and the drilled hole. At the surface, the mixture of mud and cuttings is separated so that the liquid mud is recirculated back to the well for reuse. The drill cuttings are stored in a reserve pit for later disposal. There is a growing trend to move away from reserve pits and used closed-loop drilling systems with tanks to minimize surface and ground water impacts.

The Ground Water Protection Council and ALL Consulting (GWPC and ALL 2009) provide estimates of water requirements for four of the major shale gas plays. The water required for *drilling* a typical shale gas well is approximately 3,800 m^3 in the Haynesville Shale, 1,500 m^3 in the Barnett Shale, 225 m^3 in the Fayetteville Shale, and 300 m^3 in the Marcellus Shale. To the extent that gas companies move toward longer horizontal sections in the future, the volume of water needed for drilling is likely to increase.

Completing the Well – Perforation and Hydraulic Fracturing

A newly drilled well must be properly completed to allow natural gas to enter the well and move to the surface.

Perforation: The horizontal leg of the well running through the shale layer is surrounded by steel casing. In order to permit the natural gas to enter the well, openings must be made in the casing. The holes or perforations ("perfs") are made in the casing using small explosive charges that are lowered to the desired depth on a cable. The perfs allow gas to enter the well. After creating the perfs, the well is hydraulically fractured to allow gas to flow from the formation into the well.

Hydraulic Fracturing: Because shale gas is held within the nonporous shale rock, it is necessary to fracture the shale so that the gas has a conduit or pathway to move from the shale to a production well. The process injects water, sand, and several types of additives at very high pressure into the well. The high pressure creates small fractures in the rock that extend out as 300 meters laterally away from the well. The vertical extent varies but is related to the thickness of the shale layer.

Rather than fracturing the entire horizontal leg at one time, it is done in stages of several hundred meters in length, starting with the section of the well farthest from the vertical portion of the well. Each stage is kept under pressure for only a few hours. In recent years, the newer wells tend to have longer horizontal sections, and the subsequent fracturing work is done in more stages.

After the pressure is reduced at the end of each stage, pressure is held on the well and a plug is set to isolate that fractured interval and allow fracturing of the next stage. This process is repeated for all stages. When the well is ready to be brought on production, the plugs are drilled out. Some fraction of the water used to fracture the well flows back to the surface, but most of the sand grains remain in the rock fractures, effectively propping the fractures open and allowing the gas to move. Hydraulic fracturing is done on nearly all shale gas wells; without it, the wells could not produce sufficient natural gas to be commercially viable.

Water Needed for Hydraulic Fracturing: One of the more contentious issues surrounding shale gas development is the need for new water supplies to fracture each new well. The water required for fracturing a typical shale gas well is approximately 5 million gallons (roughly 20,000 m^3). Most of the new wells being drilled have longer horizontal sections and more stages; therefore, the current volume of water needed is probably higher than those estimates.

Many operators truck fresh water either directly to a well site or to a regional water storage impoundment. In the latter case, water is pumped from regional impoundments to the well site when the fracturing work is ready to begin. Incoming water is mixed with sand and additives in onsite tanks then is injected underground.

Although many gas companies are reusing their wastewater to make up new frac fluids, still a substantial volume of new water is needed to fracture each well. The following analysis shows estimates made for the Marcellus Shale in a hypothetical high development year. In order to estimate the total amount of water needed for the Marcellus Shale region, the volume of water required to fracture one well must be multiplied by the anticipated number of wells that will be drilled in a year. The maximum number of wells estimated for each of the states in the following section is an educated guess that may prove to be inaccurate. However, the analysis provides at least an order-of-magnitude estimate.

Table 1 shows the estimated maximum number of wells that are likely to be drilled in each state in a year. For Pennsylvania and West Virginia, the number shown represents the highest number of wells actually drilled in any of the past several years (obtained from state agency data bases) multiplied by a factor of two to extrapolate to a hypothetical maximum value. New York has not

5

drilled Marcellus Shale wells in the past few years. Therefore, the number calculated for West Virginia was used as a placeholder value for New York. A volume of 20,000 m^3 was assumed for each well.

Table 1 – Cumulative Regional Estimate of Water Needs under Hypothetical Maximum Drilling Conditions

State	Hypothetical Maximum Number of Wells Drilled in a Year	Annual Volume of Water Needed
Pennsylvania	2,908	58 million m^3/yr
West Virginia	548	11 million m^3/yr
New York	548	11 million m^3/yr
Total	4,004	80 million m^3/yr = 219,000 m^3/day

These numbers are subject to various caveats however:

- The estimates of maximum wells drilled could significantly overestimate or underestimate the actual quantity. Many factors, such as the sales price of natural gas, can influence actual drilling rates.

- The assumed maximum number in one state will not necessarily correspond to the maximum in each of the other states.

- As gas companies refine and improve their efforts to recycle and reuse flowback and produced water from wells already fractured, the water needed per well may decrease.

- Conversely, if operators drill longer horizontal wells with more frac stages, the volume per well could increase.

The total water volume must be placed in context to the total water resources available within the three-state area as well as to other existing and competing uses. The U.S. Geological Survey publishes water use estimates for the United States every five years. The most recent report (Kenny et al. 2009) reports on water use for 2005. Table 2 shows the 2005 water withdrawal volumes for New York, Pennsylvania, and West Virginia by water-use category.

Table 2 – Water Withdrawals for Marcellus Shale States – 2005 (million m^3/day)

Category	New York	Pennsylvania	West Virginia	Total
Public Supply	9.6	5.4	0.7	15.7
Domestic	0.5	0.6	0.1	1.2
Irrigation	0.2	0.1	0.0	0.3

Livestock	0.1	0.2	0.0	0.4
Aquaculture	0.2	2.0	0.2	2.4
Industrial	1.1	2.9	3.7	7.7
Mining	0.1	0.4	0.1	0.5
Thermoelectric	27.0	24.4	13.4	64.8
Total	39.0	35.9	18.2	93.1

Source: Kenny et al. (2009)

Table 3 compares the estimated future water withdrawals for shale gas production with the 2005 actual water withdrawals from Kenny et al. (2009).

Table 3 – Comparison of Water Needed for Shale Gas and Total Existing Water Withdrawals

	Volume	Percentage Water Required for Shale Gas Production Compared to Total Withdrawal
Water needed for shale gas	219,000 m^3/day	-
Total water withdrawal	93.1 million m^3/day	0.24%

Although the daily volume of water needed for shale gas purposes is substantial, when compared to the existing withdrawals of water for other purposes, it is just a fraction of one percent. Similar calculations were performed for the Fayetteville Shale. The resulting percentage there was about 0.6%.

This suggests that at least from a regional perspective, the water requirements for Marcellus Shale and Fayetteville Shale gas development can be accommodated. However, in local watersheds, the incremental water demand for gas wells may create supply concerns. Water may not be available in every location or on every stream tributary, nor will it be available during every week of the year. To ensure a sustainable and available supply, the gas companies must use good advanced planning to withdraw water from rivers when flows are high and store the water until needed for fracturing. Typically this requires the construction of local or regional fresh water storage impoundments.

The two large shale gas plays in Texas appear to have different water availability trends. The Texas Water Development Board has sponsored studies in both the Barnett Shale (Bene et al. 2007) and the Eagle Ford Shale (Nicot et al. 2011). The Barnett Shale appears to have adequate available water for the time being. However, under the high demand scenario in Bene et al., groundwater resources may not be adequate to meet all Barnett Shale needs. For the Eagle Ford Shale, less information is available since it is a newer play. The local climate is somewhat drier than in the north Texas regions where the Barnett Shale is located. There is some potential for future fresh water shortages.

Gas Production and Water Management

After hydraulic fracturing is completed, and the pressure is reduced in the well, the natural gas begins flowing. It is metered at the wellhead and then sent through collection lines to a regional gas main in producing fields.

Some portion of the original frac fluids will return to the surface during the first few days to weeks. This water is referred to as flowback or flowback water. Over a much longer period of time, additional water that is naturally present in the formation (i.e., produced water) continues to flow from the well. Both flowback and produced water can contain very high levels of total dissolved solids (TDS), metals, in some cases naturally occurring radioactive materials (NORM), and other constituents. Most of these constituents were part of the shale rock and became dissolved in the flowback water while it resided in the formation.

Over an extended period of time, the volume of produced water from a given well decreases. Over this same time period, the concentration of many of the constituents continues to rise. There are few published data available that show the increase in chemical concentrations over time. One example is a report by several authors at GE (Acharya et al. 2011). They evaluated the characteristics of flowback water from the Woodford Shale. Their data show that the rates of flowback dropped from a peak of 380 to 570 liters per minute at the beginning of flowback to 75 to 150 liters per minute after 15 days. Total dissolved solids concentrations raised from 1,000 to 5,000 ppm at day 0 to 7,000 to 35,000 ppm at day 15.

Another source of TDS time-series data is a report by the Gas Technology Institute (Hayes 2009). In this case, the data come from 19 wells in the Marcellus Shale that were sampled at days 0, 1, 5, and 14. For five of the wells, additional TDS samples were taken at day 90. At day 0, all but one of the wells showed TDS concentrations less than 7,100 ppm (most were less than 1,000 ppm). However, by day 5, all TDS samples were at least 38,000 ppm (and most were above 50,000 ppm). By day 14, most samples showed TDS of more than 100,000 ppm.

The flowback volume can be 2,000 m^3 or more. Operators must be prepared to capture and store this volume as soon as the frac job is completed. In some cases, the flowback is sent to a lined pit that is used to store the flowback for several weeks until it can be disposed or reused. Other companies have chosen to eliminate flowback pits. They pump all flowback from a well to a series of dedicated tanks.

After the main volume of flowback has left the well, there is often a low steady volume of produced water that exits the well for many months. Because the volume of this flow is considerably smaller, it can be captured and stored in a tank located at each well site. Whenever the tank becomes filled, the company will arrange to have a truck remove the collected water and haul it offsite for wastewater management.

Not all of the water used to fracture a well returns to the surface as flowback water. Each shale play shows different water retention characteristics. Mantell (2011) categorizes the Barnett Shale as having formation characteristics that result in high produced water generation. As a result, the long-term produced water flow volume from the Barnett Shale is higher than most of the other shale plays. Mantell considers the Eagle Ford and Haynesville Shales to generate moderate long-term volumes of produced water. The Marcellus Shale, however, is a "very dry" shale that retains much of the water that was introduced as frac fluid. Relatively little long-term produced water flows from Marcellus Shale wells.

According to a testimony made by the Deputy Executive Director & Counsel of the Susquehanna River Basin Commission to the U.S. Senate (Beauduy 2011), only 5-12% of the original frac fluid volume from Marcellus Shale wells returns to the surface as flowback water during the first 30 days. These data are based on information from 654 wells hydraulically fractured in the Marcellus Shale during the period of October 1, 2010 to September 30, 2011.

MANAGEMENT OF FLOWBACK AND PRODUCED WATER

Operators must manage the flowback and produced water in a cost-effective manner that complies with regulatory requirements. Most of the flowback and produced water from U.S. shale gas wells is managed in one of the following ways:

- injection into a disposal well

- treatment to create clean brine

- treatment to create clean fresh water

- evaporation or crystallization

- filtration of flowback to remove suspended solids, then blend with new fresh water for future frac fluid.

These are discussed in more detail below.

Disposal Well

When shale gas development began in the Barnett Shale, operators looked to the existing and readily available technology for managing wastewater – underground injection via disposal wells. There are approximately 50,000 permitted Class II injection wells (wells are used for injection of brines and other fluids associated with oil and gas production) in Texas. About three quarters of the injection wells are used to inject water into producing formations to enhance future recovery of oil. The remaining one quarter of the wells are used to inject fluids into non-producing formations solely for disposal.

Numerous businesses operate commercial disposal wells throughout the major oil and gas fields in Texas. Tank trucks come to well sites, collect wastewater, and drive to a nearby disposal well. Wastewater is delivered by truck and unloaded into holding tanks. When a sufficient volume of wastewater has accumulated in the tanks, it is pumped through the wellhead deep underground to a porous formation.

Injection wells offer several advantages, which lead producers to favor them where possible:

- They are relatively inexpensive.

- They can be located nearby to many shale gas plays.

- Regulators are familiar with wastewater disposal in injection wells.

- Operators understand this tried and true technology.

While injection wells are heavily used in many shale gas plays, they are not used universally throughout the shale gas industry. There are two primary reasons for this. First, not every shale gas region is underlain by geological formations that will accept sufficient volumes of injected wastewater. The best-documented example of this situation is the Marcellus Shale in Pennsylvania. Operators made some early efforts to find good injection zones underneath the Pennsylvania Marcellus, but were unable to find them. This led operators to explore other wastewater management options.

Second, in some locations, fresh water is limited in availability. Operators may have difficulty obtaining sufficient quantities of fresh water for all of their future frac fluid needs. This situation provides incentives to operators to use wastewater technologies that can treat and reuse the water, thereby augmenting the available water supplies.

A third driver that may come into play as shale gas moves beyond North America is fear over impacts of underground injection. In countries that have less familiarity and comfort level with injection wells, additional restrictions may be imposed on injection wells that change the economics and potential liability that companies would face.

Treatment to Create Clean Brine

Flowback and produced water contain high levels of TDS, plus other constituents, like metals. In some locations, treatment to remove TDS is not necessary. As long as the metals are removed, the salty wastewater can be discharged or reused. There are at least two groups of technologies that are used to treat to remove the metals. One of the most common approaches used to remove metals from produced water is to raise the pH, add a coagulant or flocculant chemical to promote solids formation, and then use clarification to remove the resulting metals solids. In some instances, this is the only form of treatment used, while in other instances it serves as a pretreatment step before moving to a more advanced form of treatment.

Several small industrial wastewater treatment plants using these technologies operated for many years in Pennsylvania. These plants were designed to remove metals from natural gas wastewater and discharge a clean brine solution either to local streams under the conditions of a National Pollutant Discharge Elimination System (NPDES) permit issued by the Pennsylvania Department of Environmental Protection (PADEP) or to a municipal sewer system. Veil (2010) describes site visits to four of these plants, including the treatment steps, chemicals used, and how solids removed in the process are managed. That report provides many photos of each of the four facilities. Three of the four facilities employ only the pH adjustment, flocculation, and clarification steps. The fourth facility, Eureka Resources in Williamsport, Pennsylvania, also added an optional extra stage of treatment. This extra stage is thermal distillation that converts the clean brine into clean fresh water. However, under pressure from the PADEP, nearly all Marcellus Shale gas producers stopped sending wastewater to these small treatment facilities as of May 2011.

A second process used to create very clean brine employs a completely different technology. Ecosphere Technologies, Inc. has commercialized an advanced oxidation process (Ozonix®) that has been used treat flowback and produced water. The Ozonix® technology combines ozone generation, cavitation, and electro-chemical decomposition in a reaction vessel. The process provides near-instantaneous high-temperature oxidation that kills most microbial organisms and oxidizes metals and organics.

The company claims that their process reduces the use of biocides, scale inhibitors, and friction reducers when the treated water is reused for fracturing a new well. Ecosphere has treated flowback and produced water in the Woodford Shale and the Fayetteville Shale plays. In the only published report of Ozonix® performance, Horn (2009) describes how the Ozonix® process was used to treat flowback water from the Woodford Shale in Oklahoma during a 2008 pilot project.

Treatment to Create Clean Fresh Water

Because of the high levels of TDS in flowback and produced water, these wastewaters are particularly challenging to treat to make fresh water. Acharya et al. (2011) report on efforts to develop a wastewater treatment process for the first few hours/days of the flowback cycle that still contains relatively low levels of TDS. The TDS cutoff for consideration in the project was 35,000 to 45,000 mg/L, which is the typical limit for economic water recovery employing reverse osmosis (RO)-type membrane desalination processes. The authors concluded that membrane systems in combination with appropriate pretreatment technologies can provide cost-effective recovery of low-TDS flow-back water for either beneficial reuse or safe surface discharge.

However, when the wastewater has TDS levels higher than approximately 40,000 mg/L, nearly all of those technologies, including RO, either cannot function or they become uneconomical to use. The one technology that is able to treat high-TDS wastewater is thermal distillation. Many companies offer their own versions of thermal distillation using several different mechanical processes. The basic principle of thermal distillation is to heat the wastewater to form water vapor. The water vapor is condensed or distilled creating a clean water stream and a concentrated brine stream.

Veil (2008) describes two of the thermal processes that have been used to treat shale gas flowback and produced water. The Aqua-Pure process uses large units called NOMADs to treat wastewater. The NOMADs are mobile to the extent that they can be picked up and moved. However, relocation is a rather involved process. They can be used in centralized treatment plants for an extended period of time then they can be relocated to another location. Devon Energy did extensive testing with Aqua-Pure in the Barnett Shale. In 2010, two NOMAD units were moved from Texas to Pennsylvania for installation as an additional treatment step at the Eureka Resources wastewater treatment facility in Williamsport, Pennsylvania (Veil 2010).

The AltelaRain process is based on relatively small treatment modules. Individual AltelaRain modules are capable of processing produced water at a rate of about 1.3 m^3 per day. In order to treat larger volumes of wastewater, multiple units, run in parallel, are employed inside of large shipping containers. Like the NOMADs, the AltelaRain system can be portable or can be installed in a permanent treatment plant setup.

The Water Desalination Report (2011) notes that four new AltelaRain modules were installed in a new centralized wastewater treatment plant (owned and operated by Clean Streams LLC) in Williamsport, Pennsylvania during 2010. The plant can treat approximately 400 m^3 of produced and flowback water per day.

According to a report describing a long-term pilot test of an AltelaRain system in Pennsylvania (Bruff and Godshall 2011), the total operating costs, including disposal of the brine, were $0.03/liter ($4.53/bbl). This represents a savings of 28% over trucking and disposal without the AltelaRain System. Additionally, in many cases throughout the Marcellus Shale, costs exceed $0.04/liter ($6.30/bbl) including trucking.

Other companies offering thermal treatment processes are also trying to get a foothold in the competitive shale gas market. The cost to operate thermal treatment systems makes them less attractive than some other options in many locations.

Evaporation/Crystallization

The EVRAS (Evaporative Reduction and Solidification) system evaporates the entire volume of flowback and produced water, and the solids are crystallized. Chesapeake Energy uses an EVRAS system at its gas processing plant in Fort Worth, Texas. The system utilizes waste heat from the gas processing plant. However, in the absence of a free or inexpensive source of heat or energy, evaporation systems become very expensive. The advantage they offer is that there are no liquid byproduct streams.

Other companies offer "zero liquid discharge" or crystallizations technologies, but no further published reports were found to document their actual use on shale gas flowback and produced water.

Treatment to Remove Solids and Reuse the Wastewater in Future Frac Fluids

One offshoot of the lack of a good injection formation beneath the Pennsylvania Marcellus Shale is that operators were denied their preferred wastewater management option (injection) and were forced to innovate. Veil (2010) reports on interviews with Marcellus gas operators in spring 2010. At that time, the larger companies were beginning to experiment with capturing their flowback water, passing it through a simple filter to remove sand grains and scale particles, then blending it with fresh water to make up frac fluids for a future well. This was a novel process at the time, because many thought that the high concentrations of TDS, metals, and other constituents would interfere with the performance of the frac fluid in the next well.

Range Resources, one of the pioneers of this approach, found that frac fluids that include some recycled flowback get production results that are comparable to those fractured with all fresh water. Range Resources had no indication of issues with frac fluid stability, scaling, or downhole bacterial growth. During 2009, Range Resources completed 44 wells and did frac jobs involving 364 stages. The total volume of frac fluid used was 600,000 m^3, with 28% of the volume made up of recycled water from a previous well (Gaudlip 2010).

By late 2010, several other large gas companies in the Marcellus Shale region were actively utilizing filtration and reuse for all or most of their flowback water.

Differences in Managing Flowback Water vs. Produced Water

It is interesting to note that even though much of the initial Marcellus flowback is reused for new frac fluids following filtration, relatively little of the ongoing low-volume produced water is managed in that way. Most wells have one or more water tanks onsite. The produced water accumulates in tanks at every well. Every week or so, a vacuum truck comes by and collects the water and takes it to a disposal or treatment facility.

One possible explanation for this dichotomy in wastewater management approach is that it is not too difficult or costly to collect and treat a substantial volume of water at a single location at one time. However, the complexity and cost increases when an operator tries to collect modest volumes of water from dozens to hundreds of locations spread over several counties and treat that water for reuse. In addition, rather than being a one-time operation, the produced water continues indefinitely from each well (old and new alike).

Theoretically, a mobile treatment device that could visit a site and process a tank of produced water onsite in a short amount of time could have some applicability. However, once the water is treated onsite, there are two byproduct streams that must be managed in some way. For the sake of example, assume a tank holding 20 m^3 of produced water treated onsite with an 80% treatment efficiency. This would result in 16 m^3 of clean water (note that under federal regulations, that water cannot be discharged onsite, so what can be done with it?) and 4 m^3 of concentrated brine that must be trucked offsite for disposal. If there are hundreds of wells in a county, the logistics of moving the clean water and concentrated brine become costly and complicated.

CHEMICAL COMPOSITION OF FRAC FLUIDS

Hydraulic fracturing is used to create a network of cracks in an underground formation. The frac fluids are injected in a controlled process under high-rate, high-pressure conditions. The most common type of fracturing used in shale gas formation is called a "slickwater frac". The fluids used for this type of frac job consist of approximately 90% water, 10% sand or some other proppant that helps to keep the newly created fractures open after the pressure is released, plus a variety of other chemicals added in very low concentrations for control purposes. For example:

- biocides are used to control microbial growth in the formation that could lead to hydrogen sulfide production,
- corrosion inhibitors are used to prevent corrosion in the well casing and pipes, and
- gel is used to thicken the frac fluid so that the sand proppant can remain suspended and move far out into the fractures.

GWPC and ALL (2009) includes several useful tables and figures that describe the types of chemicals used and their relative proportions in the final frac fluid. Figure 3 is a graphic reproduced from that report. It shows that the combined total of 12 types of chemical additives makes up less than one half of one percent of the total frac fluid volume.

13

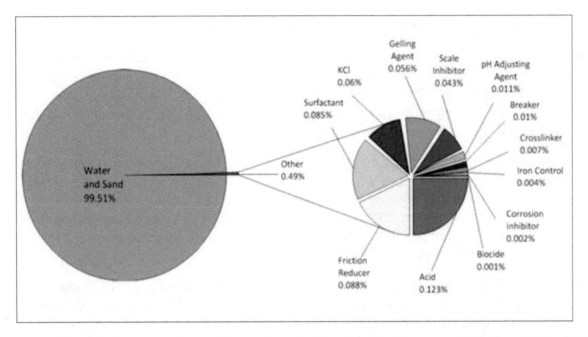

Figure 3. Relative Proportions of Ingredients in Frac Fluids. Source: GWPC and ALL (2009); note that this report was funded by the U.S. government, so it is in the public domain.

Diclosure of Chemical Used in Frac Fluids

One of the most contentious issues surrounding hydraulic fracturing is that neither the oil and gas companies nor the large service companies that perform the frac jobs have historically shared detailed information with regulators or the public on which chemicals are actually used in frac jobs. Opponents of hydraulic fracturing often portray the frac fluids as being heavily laden with toxic chemicals. While some chemicals with toxic properties are used, the concentrations are low. But even if the chemicals used are not harmful, the public has concerns over the unknown and does not trust the industry to safeguard them.

The type of information that is typically desired by the public and the media about each chemical are:

- the trade name (often additives are given a product name that includes letters and numbers but does not really explain what is in the product -- e.g., Ajax 356P),
- what ingredients are really in it (e.g., polymer, water, mineral spirits)
- why it is used (e.g., to prevent scale formation)
- how much is used in that well
- total quantity (volume and/or mass)
- concentration (ppm, mg/l)
- is it harmful to human health?

MSDSs: Much of this information can be found in documents called Material Safety Data Sheets (MSDSs) that are required by the U.S. Occupational Safety and Health

14

Administration for any product that contains hazardous or carcinogenic materials of specific threshold values (1.0% and 0.1% respectively) but not for all chemical products. However, the MSDSs do not contain all the desired information and are only useful when the names of the products have been identified first.

Frac Focus: In an effort to help provide more information to the public on chemical usage in individual wells, two organizations representing state agencies – the GWPC and the Interstate Oil and Gas Compact Commission (IOGCC) – developed an online Chemical Disclosure Registry system that is available at the Frac Focus website (www.fracfocus.org). The Registry was launched in April 2011, and gas companies were asked to voluntarily enter the details of the chemicals used for each hydraulic fracturing job. The initial response from the industry was quite good. Over the next year, several states adopted chemical disclosure regulations, and many of those required that operators enter the data into the Frac Focus website. The success of this program can be measured by the amount of information already added to the Registry. As of April 30, 2012 (just over one year after the Registry was opened), data had been entered on more than 16,000 wells representing 129 gas companies.

Interested viewers can visit the site and search for a well using any of the following sorting factors:

- state
- county
- operator
- well name
- API well number.

Figure 4 shows a sample output screen from Frac Focus. The data are from Chesapeake Resources Well BSOA 14-14-15 H-1, located in De Soto County, Louisiana. The well was hydraulically fractured on March 21, 2011. It is one of the earliest wells entered into the Registry.

The first three columns show the trade name of each additive, who supplied it, and why it is used. The fourth and fifth columns list the ingredients and Chemical Abstract number for each additive. The sixth and seventh columns show the percentage by mass of each ingredient within the additive product (6th column) and within the full mixed frac fluid (7th column). Note that other than water and sand, all the other ingredients are small fractions of one percent.

Trade Name	Supplier	Purpose	Ingredients	Chemical Abstract Service Number (CAS #)	Maximum Ingredient Concentration in Additive (% by Mass)**	Maximum Ingredient Concentration in HF Fluid (% by Mass)**
Fresh Water		Carrier/Base Fluid				86.12803%
Sand (Proppant)		Proppant				12.83614%
Acid, 15% HCl	CUDD ENERGY SERVICES	Acid	Water	007732-18-5	85.00%	0.06070%
			Hydrochloric Acid	007647-01-0	15.00%	0.01071%
I-22	CUDD ENERGY SERVICES	Corrosion Inhibitor	Formic Acid	000064-18-6	60.00%	0.00053%
			Aromatic aldehyde	N/A	30.00%	0.00026%
			Haloalkyl heteropolycycle salt	N/A	30.00%	0.00026%
			Oxyalkylated Fatty Acid	N/A	30.00%	0.00026%
			Isopropanol	000067-63-0	5.00%	0.00004%
			Methanol	000067-56-1	5.00%	0.00004%
			Organic sulfur compound	N/A	5.00%	0.00004%
			Quaternary ammonium compound	N/A	5.00%	0.00004%
			Benzyl Chloride	000100-44-7	1.00%	0.00001%
SG-15M	CUDD ENERGY SERVICES	Gelling Agent	Petroleum Distillate	064742-47-8	55.00%	0.06860%
			Guar Gum	009000-30-0	50.00%	0.06236%
			Clay	014808-60-7	2.00%	0.00249%
			Surfactant	068439-51-0	2.00%	0.00249%
BUFFER H	CUDD ENERGY SERVICES	pH Adjusting Agent	Water	007732-18-5	94.50%	0.02070%
			Sodium Hydroxide	001310-73-2	51.50%	0.01128%
			Sodium Chloride	007647-14-5	5.00%	0.00110%
GB-4	CUDD ENERGY SERVICES	Breaker	Proprietary	N/A	100.00%	0.00120%
CX-14G	CUDD ENERGY SERVICES	Cross Linker	Petroleum Distillate Hydrotreated Light	064742-47-8	60.00%	0.01454%
GB-2	CUDD ENERGY SERVICES	Breaker	Ammonium Persulfate	007727-54-0	100.00%	0.00083%
NE-21	CUDD ENERGY SERVICES	Non-Emulsifier	Methanol	000067-56-1	30.00%	0.01218%
			Oxyalkylated alcohols	N/A	30.00%	0.01218%
			Ethoxylated Alcohols	N/A	10.00%	0.00406%
CX-14A	CUDD ENERGY SERVICES	Cross Linker	Sodium Tetraborate	001330-43-4	25.00%	0.00056%
CS-125C	CUDD ENERGY SERVICES	Clay Stabilizer	No Hazardous Components	NONE		0.00000%
FRA-4	CUDD ENERGY SERVICES	Friction Reducer	No Hazardous Components	NONE		0.00000%
MC B-8642 (WS)	MULTI-CHEM GROUP LLC	Anti-Bacterial Agent	Glutaraldehyde (Pentanediol)	000111-30-8	60.00%	0.01180%
			Quaternary Ammonium Compound	068424-85-1	10.00%	0.00197%
			Ethanol	000064-17-5	1.00%	0.00020%
MC S-2510T (WS)	MULTI-CHEM GROUP LLC	Scale Inhibitor	Ethylene Glycol	000107-21-1	60.00%	0.00605%
			Sodium Hydroxide	001310-73-2	5.00%	0.00050%

Figure 4. Sample Output Screen from Frac Focus Chemical Registry. Source: www.fracfocus.org

CONCLUSIONS

Shale gas represents a valuable form of natural gas that is available in abundant supply within the United States and elsewhere in the world. Water is involved in several steps of the shale gas development process. In particular, significant volumes of water are needed to support drilling and hydraulic fracturing. Following completion of a frac job, some of the water injected as frac fluids returns to the surface and must be managed. There are a variety of options for managing the flowback and produced water from shale gas production. Some of the factors that determine how the wastewater will be managed are:

- the volume of water generated,
- the specific chemical and physical characteristics of the wastewater,
- the practicality and feasibility of an option,
- the regulatory requirements that allow or block an option,
- the long-term liability of chosen options,
- the costs (considering all cost components) of using an option in a way that meets regulatory requirements.

Much of the flowback and produced water from U.S. shale gas operations is managed by underground injection into disposal wells. This is often the lowest cost option that meets regulatory requirements, and is therefore favored by the gas companies. But there are several circumstances under which injection may not be the favored option. First, some areas do not have good injection formations readily available. Second, in some regions, fresh water is in short supply, thereby adding value to recycling flowback. Finally some companies may preferentially choose to follow sustainable practices and voluntarily look to recycle or reuse their wastewater.

The options actually used are evolving over time. As regulations become more stringent and new technologies enter the market place, gas companies may shift from one wastewater management option to another.

REFERENCES

Acharya, H.R., C. Henderson, H. Matis, H. Kommepalli, B. Moore, and H. Wang, 2011, "Cost Effective Recovery of Low-TDS Frac Flowback Water for Re-use," prepared for the U.S. Department of Energy, National Energy Technology Laboratory by GE Global Research, June, 100 pp.

ARI, 2011, "World Shale Gas Resources: An Initial Assessment of 14 Regions Outside the United States," prepared for the Energy Information Administration by Advanced Resources International, April. Available at http://www.eia.gov/analysis/studies/worldshalegas/pdf/fullreport.pdf; accessed September 7, 2011.

Beauduy, T.W., 2011, testimony to the Subcommittee on Water and Power, Senate Committee on Energy and Natural Resources, United States Senate, Hearing on Shale Gas Production and Water Resources in the Eastern United States, October 20. Available at: http://www.srbc.net/programs/docs/Water%20&%20Power%20Subcommittee%2010_20_11%20SRBC%20Beaudy%20Testimony.pdf.

Bené, J., B. Harden, S. W. Griffin, and J.-P. Nicot, 2007, "Northern Trinity/Woodbine GAM Assessment of Groundwater Use in the Northern Trinity Aquifer Due To Urban Growth and Barnett Shale Development," prepared for the Texas Water Development Board, January. Available at http://www.twdb.state.tx.us/RWPG/rpgm_rpts/0604830613_BarnetShale.pdf.

Bruff, M., and N. Godshall, 2011, "An Integrated Water Treatment Technology Solution for Sustainable Water Resource Management in the Marcellus Shale," prepared for U.S. Department of Energy, National Energy Technology Laboratory by Altela, Inc., February 28, 188 pp.

EIA, 2012, "Annual Energy Outlook 2012, Early Release Overview," U.S. Department of Energy, Energy Information Administration, January 23, available at http://www.eia.gov/forecasts/aeo/er/.

GWPC and ALL, 2009, "Modern Shale Gas Development in the United States: A Primer," prepared by the Ground Water Protection Council and ALL Consulting for the U.S. Department of Energy, National Energy Technology Laboratory, April, 116 pp.

Gaudlip, T., 2010, "Preliminary Assessment of Marcellus Water Reuse," presented at Process-Affected Water Management Strategies conference, Calgary, Alberta, Canada, March 17.

Hayes, T., 2009, "Sampling and Analysis of Water Streams Associated with the Development of Marcellus Shale Gas," prepared for the Marcellus Shale Coalition by Gas Technology Institute, December 31, 210 pp.

Horn, A., 2009, "Breakthrough Mobile Water Treatment Converts 75% of Fracturing Flowback Fluid to Fresh Water and Lowers CO2 Emissions," SPE 121104, presented at the SPE Americas E&P Environmental and Safety Conference, San Antonio, Texas, March 23-25.

Kenny, J.F., N.L. Barber, S.S. Hutson, K.S. Linsey, J.K. Lovelace, and M.A. Maupin, 2009, "Estimated Use of Water in the United States in 2005," U.S. Geological Survey Circular 1344, 52 pp.

Mantell, M.E., 2011, "Eagle Ford Water Sourcing and Produced Water Reuse: Applying Experience from Other Shale Plays," presented at the SPE Eagle Ford Shale Workshop, Austin, Texas, August 25.

Nicot, J.-P., A. K. Hebel, S. M. Ritter, S. Walden, R. Baier, P. Galusky, J. Beach, R. Kyle, L. Symank, and C. Breton, 2011, "Current and Projected Water Use in the Texas Mining and Oil and Gas Industry," draft report prepared for the Texas Water Development Board, February, 354 pp.

Veil, J.A., 2008, "Thermal Distillation Technology for Management of Produced Water and Frac Flowback Water," Water Technology Brief #2008-1, prepared for U.S. Department of Energy, National Energy Technology Laboratory, May 13, 12 pp. Available at http://www.veilenvironmental.com/publications/pw/ANL-EVS-evaporation_technologies2.pdf.

Veil, J.A., 2010, "Water Management Technologies Used by Marcellus Shale Gas Producers," ANL/EVS/R-10/3, prepared for the U.S. Department of Energy, National Energy Technology Laboratory, July, 59 pp. Available at http://www.veilenvironmental.com/publications/pw/Water_Mgmt_in_Marcellus-final-jul10.pdf.

Water Desalination Report, 2011, "Distiller Reduces Frac Water Volume," May 9 issue, available at http://www.altelainc.com/images/uploads/2011_5-9_Water_Desalination_Report.pdf.

HYDRAULIC FRACTURING IN THE CONTEXT OF SUSTAINABLE WATER MANAGEMENT

Ethan T. Smith[1] and Harry X. Zhang[2]

[1] Sustainable Water Resources Coordinator, c/o U.S. Geological Survey (retired), Reston, Virginia 20192; Email: etsmithsiri@aol.com

[2] Principal Engineer and Industrial Water Resources Lead, CH2M HILL, 15010 Conference Center Drive, Chantilly, Virginia 20151; Email: Harry.Zhang@CH2M.com

ABSTRACT

A large amount of clean, domestic energy in the form of unconventional shale gas exists in the U.S. In the last 15 years, opportunities for obtaining shale gas have grown increasingly attractive with the evolution of horizontal drilling and hydraulic fracturing technologies, greatly escalating production nationwide. However, there exists vocal public concern over the impact of shale gas production and the chemical constituents used in hydraulic fracturing on groundwater and surface water resources, local infrastructure and air quality. Therefore, balancing energy production and mitigating negative environmental and social impact will require proactive management and cooperation. Both technical and institutional contributions are needed to achieve this delicate balance. Surface and groundwater supply can be affected by development, especially when existing sources are near capacity. Disposal of produced water can be a serious issue due to the nature of the contaminants. Site-specific analysis may be the best way to address the issues involved, as demonstrated in Fayette County, Pennsylvania. Balancing energy production with environmental impact is an evolving public policy issue that requires public and private cooperation.

KEYWORDS: Hydraulic fracturing, fracking, sustainable water management, water sustainability, shale gas, unconventional gas

INTRODUCTION

The authors have worked since 2004 to develop a holistic picture of water resources sustainability and have published a series of papers on the alternative methods to achieve this sustainability. In this paper, the concept is applied to hydraulic fracturing, because of the importance of the process to energy development in the nation and the world. The advent of gas extraction from shale deposits offers great hope toward achieving energy independence; however, the potential impacts on water and other environmental resources has called this bounty into question. Claims of the pros and cons about hydraulic fracturing vary tremendously. Apparently, more investigation is needed to determine the proper balance between obtaining this important resource and the possible negative impacts on water resources.

Achieving a sustainable balance can be better understood by looking at some simple definitions of sustainability, as illustrated in Table 1. A more complete discussion of this definition can be found on the page "What is Sustainability" on the website http://sites.google.com/site/sustainablewaterresources. A short rule of thumb might be "to what extent will the area continue in its current pattern of human use over an extended time span?"

Table 1: Key Features for Various Levels of Sustainability

Consumption of Renewable Resources	State of Environment	Level of Sustainability
More than nature's ability to replenish	Environmental degradation	Not sustainable
Equal to nature's ability to replenish	Environmental equilibrium	Steady-state sustainability
Less than nature's ability to replenish	Environmental renewal	Sustainable development

Sources: Daly (1973 and 1996); Executive Order 13423 (2007)

When faced with a new application like shale energy development, there is usually a choice between preventing adverse impacts and remediating to correct such impacts after they become apparent. In many cases the national choice has favored remediation, even though this is often the more difficult and costly option.

General discussions of hydraulic fracturing can be found from the International Water Association (http://iwawaterwiki.org/xwiki/bin/view/Articles/Hydraulicfracturing). The recent history of shale gas development in the U.S. is given by Arthur (2009) as shown:

- Hydraulic Fracturing Used in the Oil and Gas Industry (1950-1960s)
- Barnett Shale – Ft. Worth Basin Development (1982)
- Horizontal Wells in Ohio Shales (1980s)
- Successful Horizontal Drilling in Barnett Shale (2003)
- Horizontal Drilling Technology Applied in Appalachian Basin, Ohio, and Marcellus Shales (2006)

Horizontal drilling and hydraulic fracturing are key technologies in the economic success of modern shale gas development. This has not come without cost, however. Arthur (2009) also discusses the water management issues that accompany shale gas development: Drilling and hydraulic fracturing can require between 2 and 5 million gallons of water per well, competing with existing human, industrial/agricultural and environmental uses in some areas. The temporary, geographically dispersed and often undeveloped nature of many drilling and completion operations complicates logistics of delivering water to and transporting waste from work sites, even in areas of the nation regarded as being well supplied. To supplement water supply, fracturing fluids that can use lower-quality water, such as surface water, groundwater, or industrial wastewater with higher total dissolved solids (TDS) would be very desirable.

Water produced as a result of the fracturing process must be treated, reused, or disposed. The flowback water may initially be similar in quality to the source water used for hydraulic fracturing; and containing small amounts (generally less than 1%) of the chemicals used in the process. As flowback continues, the TDS in the water increases as the produced water becomes more and more influenced by naturally occurring geology and formation water, in some regions approaching 5 to 7 times the salinity of seawater. This chemical nature can limit the options for disposal, along with the volume of water to be treated. Constituents of concern in flowback and produced water are not addressed in typical municipal wastewater treatment processes, an early practice, and can require sophisticated and expensive water treatment technology to render these waste streams safe for discharge. The most common and cost effective disposal strategy is deep well injection to a confined formation that is isolated from potential groundwater aquifers that could be used for potable water supply.

METHODOLOGY AND RESULTS

Shale resources exist on a national basis, as shown in Figure 1. Figure 2 shows gas production for different regions in the U.S. However, for the purposes of this study, attention will be focused primarily on the Marcellus Shale, with specific attention to the state of Pennsylvania (Vidic, 2010). In the Marcellus Shale, there is concern about possible interaction between produced well water and surrounding groundwater supplies, leading to contamination of the water supply with gas and other chemicals. The substantial difference in depth between the shale formations, as compared to local ground water supply, would seem to raise doubts as to the possibility of interference. Yet, concerns have led to questions about pathways from leaking well casings from historic oil production in region over eth past century and from poorly constructed/maintained private water supply wells and other causes. The movie *GasLand* includes powerful images of lighting tap water with a match.

The constituents of the fluids used in the fracturing process raise questions, mostly because of the possibility of groundwater contamination. Additives can escape during mixing or spills, or remain in the fluids recovered after injection of the fluid. Vidic (2010) gives a list of typical constituents, as listed in Table 2. Over the past several years, there have been several voluntary and regulatory-mandated programs for companies to disclose the formulas of additives to address public concerns.

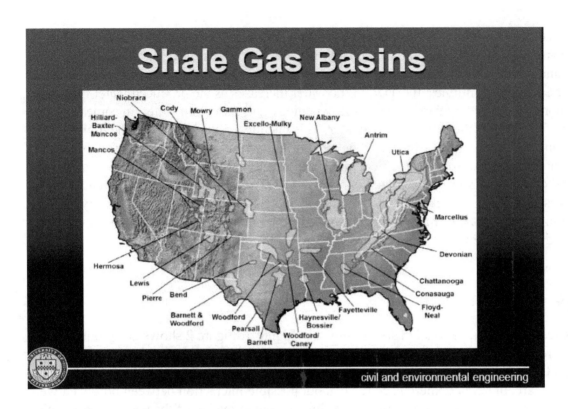

Figure 1: Shale Gas Basins of the United States (Vidic, 2010)

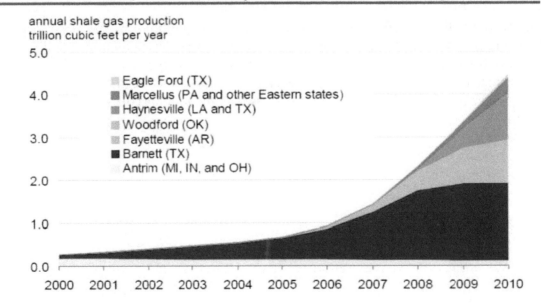

Table 2: A List of Typical Constituents in Fracking Fluids

Additive type	Main Compound	Purpose
Diluted acid (15%)	Hydrochloric or Muriatic	Dissolve minerals and initiates cracks in rock
Biocide	Glutaraldehyde	Bacterial control
Corrosion inhibitor	N,n-dimethyl formamide	Prevents corrosion
Breaker	Ammonium persulfate	Delays breakdown of gel polymers
Crosslinker	Borate salts	Maintains fluid viscosity at high temperature
Friction reducers	Polyacrylamide	Minimize friction between the fluid and the pipe
	Mineral oil	
Gel	Guar gum or hydroxyethyl cellulose	Thickens water to suspend the sand
Iron control	Citric acid	Prevent precipitation of metal oxides
Oxygen scavenger	Ammonium bisulfite	Remove oxygen form fluid to reduce pipe corrosion
pH adjustment	Potassium or sodium carbonate	Maintains effectiveness of other compounds (e.g., crosslinker)
Proppant	Silica quartz sand	Keeps fractures open
Scale inhibitor	Ethylene glycol	Reduce deposition on pipe
Surfactant	Isopropanol	Increase viscosity of fluid

Although estimates of water use for shale gas production are only about 1 percent of the watershed supply, there may still be short-term impacts on other water uses, such as public supply, water-based recreation, and in-stream environmental needs (Panko, 2011).

It is important to have a general picture of how the hydraulic fracturing process works (see Figure 3). The localized nature of the process immediately raises the question of how to analyze this kind of situation. The national-level water indicators that have been used elsewhere are too gross for such site-specific situations. From this point on, local-level water indicators of sustainable development offer a much better chance in a quest to determine how best to balance energy production versus environmental impact.

What Can Be Done?

Some general technical methods can be used to mitigate the impacts of fracturing (Arthur, 2009). To obtain water supply, local surface water and groundwater might be used, as well as water from municipal supplies and wastewater treatment plants. The produced water itself might be reused, unless the TDS and chloride concentrations are too high. The produced water might be treated, but there are limitations due to quality. When the quality becomes too poor, perhaps only 50 percent of the quantity can be reused (an approximate practical recovery limit for the most aggressive treatment technologies on these types of streams) , leaving the remainder for disposal. Disposal might be in an injection well (there are limited opportunities for this in the Marcellus Shale), or treatment at a municipal or industrial treatment facility (there may be few that can treat this type of waste). At a detailed level, the Citizens' Guide to Marcellus Shale in Pennsylvania, 2010 (see http://seagrant.psu.edu/news/marcellus_citizens_guide.pdf) offers a variety of ways to cope with the problem. The Guide takes the user through the entire process, including initial permitting, wastewater, drinking water contamination, noise and esthetic impacts, land and forest impacts, and air quality. Many helpful links to additional information are included.

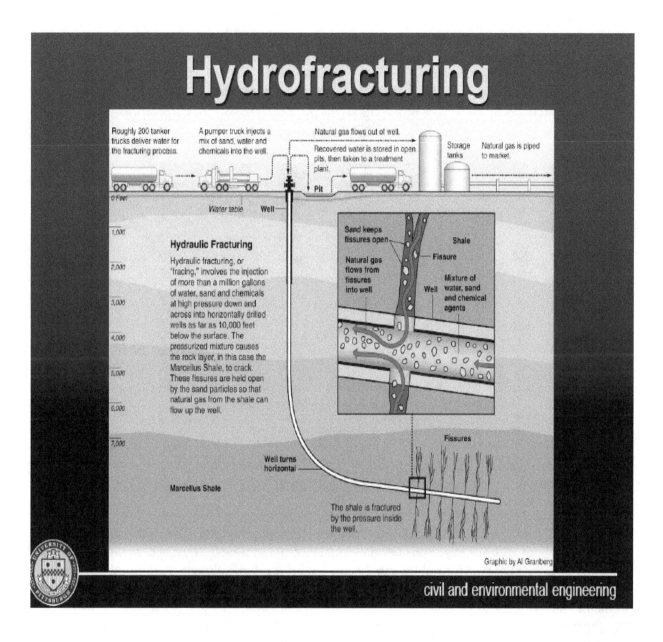

Figure 3: The Hydraulic Fracturing Process (Vidic, 2010).

Site-Specific Evaluation

To reach a more site-specific level, we can go to a more detailed geographical scale and examine conditions in Fayette County, Pennsylvania, which is located in the far southwestern part of the state. This is an active producing area and has some good data. Figure 4 shows some of this information in terms of water withdrawals, wastewater treatment plants, and shale gas wells in the state.

27

Water Withdrawals, Wastewater Treatment Facilities and Marcellus Wells

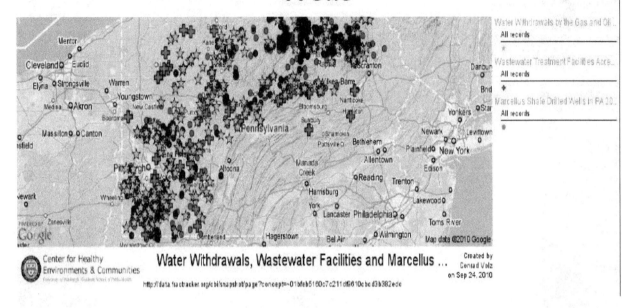

Figure 4: Pennsylvania Water Withdrawals, Wastewater Facilities, and Marcellus Wells. (Volz, 2011). See http://energy.wilkes.edu/PDFFiles/Issues/Volz_Jan2011_ProgressSummit.pdf

Water use in the Marcellus Shale area is dominated by power generation, public supply, and industrial and mining uses (Arthur, 2009). In Fayette County, the total 2005 water withdrawal for all uses was about 43 million gallons per day (mgd) (http://water.usgs.gov/watuse). Of this amount, about 35 mgd was for public supply and 8 mgd for domestic use. Insights about water supply can be found by looking at the United States Geological Survey (USGS) stations shown on the interactive map at http://waterdata.usgs.gov/pa/nwis/inventory?county_cd=42051&format=sites_mapper&sort_key =site_no&group_key=NONE&sitefile_output_format=html_table&column_name=agency_cd&c olumn_name=site_no&column_name=station_nm&list_of_search_criteria=county_cd.

The major rivers for this county are the Monongahela and the Youghiogheny, if we examine only surface water availability. For the Monongahela River, the discharge in Fayette County is in the range of about 2,000 cubic feet per second (cfs) (1,293 mgd) to 35,000 cfs (22,621 mgd). For the Youghiogheny River, the discharge is about 2,000 cfs (1,293 mgd) to 5,000 cfs (3,232 mgd). The groundwater recharge estimates for the county are about 10 to 14 inches per year (Reese, 2010).

A typical shale well might use about 2 to 5 million gallons per well. Even if we ignore the possibilities of water reuse and assume that all water is acquired from readily available local surface-water sources, a simple calculation shows that 5 million gallons per well can be readily accommodated. Looking at the above supply from the local rivers, there appears to be ample water in the county. In many cases, the water might be delivered to the well site by tank truck, which is one of the community impacts often mentioned in these rural areas.

Sumi (2008) indicates that in 2008 there were 58 wells and 250,000 leased acres in the county. Other sources imply that this total is much greater. Figure 4 shows a good deal of development for shale wells, along with the associated water and wastewater facilities. It also shows facilities that may be designed to treat wastewater from the hydraulic fracturing process. Questions remain about how well this arrangement is working. The wastewater from drilling operations is handled poorly by municipal facilities that are designed for biological waste.

Besides the water availability challenge, water quality also remains an issue of concern in the Monongahela River watershed. On May 1, 2010, the Pennsylvania Department of Environment Protection (PADEP) issued a guidance document entitled "Coordinating National Pollutant Discharge Elimination System (NPDES) Permitting in the Monongahela River Watershed." This document provides evaluation criteria to PADEP staff for use in evaluating permit applications for facilities with the potential to discharge the pollutants of concern identified as exceeding water quality standards, i.e., TDS and sulfate. This guidance applies to permit applications for any new dischargers, new sources, and expanding facilities that discharge TDS or sulfate directly or indirectly (via a tributary) to the Monongahela River where DEP determines that effluent limits are necessary to achieve water quality standards.

Local stakeholders created the Marcellus Shale Task Force of Fayette County (see http://www.mshaletaskforce.org) to address perceived problems. The website is one source of data on active wells, site locations, and gas exploration companies. Seeking a balance between energy production and the regional economic growth that can result, particularly in areas that have been historically depressed for long period, and environmental impact is an evolving public policy issue and will require more effort from public and private entities to reach an acceptable compromise.

Assessing Sustainability

Looking at all the information available, it remains to be seen how the hydraulic fracturing process relates to the idea of water sustainability. It must be admitted that we are at an early stage of the process, in which it is difficult to make an unequivocal estimate of sustainability. So, what is to be done? Perhaps one initial approach would be to use the system definitions shown on the "What is Sustainability" page at http://sites.google.com/site/sustainablewaterresources/, and then to read the hydraulic fracturing case study in Fayette County. Following are the results of such an attempt:

Duration

Figure 2 shows that production from shale gas formations can last for many years. This implies that long-term change from prior land uses may occur. However, after drilling and completion and once the well is placed into production, only a small installation will remain on the land where the produced gas is extracted. Consequently, agricultural activity would be minimally affected. Wells also seem to have a life cycle of their own. Swackhamer (2012) states that a given well has a productive lifetime on the order of 5 to 7.5 years. Based on this information, it appears that shale gas production would not be a short-term change, but rather meets the definition for long-term duration.

Rate of Change

The Pennsylvania experience shows that shale gas production has occurred rapidly over a relatively short span of years, giving the system of institutions little time to react in a thoughtful fashion. Time allowed for communities, infrastructure and regulations and the agencies themselves to adapt has been minimal, and the system is in a classic case of system shock. Not surprisingly, there has been a backlash of local institutions, with an accompanying list of allegations and counter-allegations by opposing parties. Perhaps mistakes have occurred in drilling technology, since this is an early-adopter area. How can the volume of waste chemicals be handled, since conventional wastewater treatment may not be very effective? For example, the produced water may be 10 times more saline than seawater (Jackson, 2012). Spills, upward migration of gas, and explosions may occur (Swackhammer, 2012). What about the impact of imposing an industrial technology on rural agricultural communities? A long list of such problems is listed by Vidic (2010). The conclusion based on how rapid change has impacted the system is that the resulting instability indicates a lack of sustainability.

Dynamic System Properties, Buffering, and Probability

To date, there has not much sign of stability in how this institutional system behaves. Transformation has greatly exceeded the system's ability to absorb change without effect. If only a few wells existed, the character of the system might continue in its earlier mode. Public outcry would suggest that shale gas development has evidently exceeded buffering capacity. However, there is not enough data to assess the probability of stochastic system elements; that is, we cannot use probability to predict what the system might do next. All of these characteristics show a lack of sustainability. Another possibility is that the system has passed a "tipping point," in which it will flip from an agricultural area to an industrial area. In that case, the nature of the system will change, and we will experience a completely different system.

Reversibility and Extreme Conditions

One of the properties of a sustainable system is the ability to reverse actions if we encounter unexpected adverse outcomes. Like many other interventions that humans have carried into the environment, hydraulic fracturing shows very limited opportunities for such reverse actions. Setting aside the mounting demand for the energy resource itself, the land use is changed dramatically once drilling is conducted. Not only are the drill sites themselves greatly different from any pre-existing agricultural land uses, but many comments have surfaced about community or regional impacts too. To service drill sites, roads must be heavily used, and sometimes new access roads must be built. The industrial character of drilling requires trucking of water and wastewater, and perhaps construction of storage tanks or lagoons. If available wastewater treatment facilities are unsuitable, specialized industrial wastewater facilities must be constructed. All of these activities can be expected to have sociological and economic effects in the region. The aggregate of all these effects would argue for a lack of sustainability in the hydraulic fracturing process.

Extreme conditions are often an indication of an unsustainable system. If variables are forced to limits such as zero or one, we are alerted to the fact that something is wrong. Do such effects occur in hydraulic fracturing? If we examine the spatial density of wells, the answer may be that they do. For example, if there are only a few wells, they may have minimal impact. Evidently, a great deal depends on how many wells are drilled per spatial unit. Sumi (2008) includes some data on such wells. Although state regulations initially may require greater spacing, there are cases in which wells are being drilled about 250 feet to 500 feet apart, in the Barnett Shale. Looking at other formations nationwide, spacing on the order of 40 acres to 160 acres is found. Regulations in West Virginia require 1,000 feet minimum spacing. The pattern is complex, but the trend is toward a tighter spacing of wells, depending on the economics of the resource. This trend is exactly the sort of push toward an extreme condition that may well indicate a lack of sustainability. Therefore, a smaller and more dispersed surface footprint will be a good practice towards sustainability.

SUMMARY AND CONCLUSIONS

The projected demand for energy is expected to increase substantially, making it necessary to use the energy resources that apply the best technology available. This will necessitate tradeoffs among a number of variables to achieve an acceptable balance between energy production and environmental impact. Finding this delicate balance will require both technical and institutional contributions.

Surface water supply and groundwater supply can both be affected by development, especially when existing availability is at or near capacity. Disposal of produced water from well production can be very difficult, primarily due to the nature of the contaminants. Many of these contaminants are currently not treatable at most facilities designed to treat secondary wastewater. Site-specific analysis seems to be the best way to address the issues involved. Seeking a balance between energy production and environmental impact is an evolving public policy issue and will require more effort from public and private entities to reach an acceptable compromise in future years.

ACKNOWLEDGMENTS

The authors would also like to express their thanks for comments and suggestions received from reviewers during the review process.

REFERENCES

Arthur, J. Daniel. January 2009. Prudent and Sustainable Water Management and Disposal Alternatives   Applicable to Shale Gas Development, Ground Water Protection Council. San Antonio, Texas. http://www.energyindepth.org/PDF/ALL-Shale-Gas-Water.pdf.

Daly, H. E. 1973. Toward a Steady State Economy. Freeman: San Francisco, California.

Daly, H. E. 1996. Beyond Growth: the Economics of Sustainable Development. Beacon Press: Boston, Massachusetts.

Engelder T., and Lash, G.G. 2008, Marcellus shale play's vast resource potential creating stir in

Appalachia: American Oil and Gas Reporter, 51(6), p76-87.

Executive Order 13423. 2007. Strengthening Federal Environmental, Energy, and Transportation Management. Federal Registrar 2007, 72 (17), 3919–3923.

Focazio, M., 2012. The Geographic Footprint. USGS. Workshop on the Health Impact Assessment of New Energy Sources: Shale Gas Extraction, National Academy of Science. http://www.iom.edu/frackingHIA.

Jackson, R. 2012. Hydraulic Fracturing, Water Resources, and Human Health. Duke University Workshop on the Health Impact Assessment of New Energy Sources: Shale Gas Extraction. National Academy of Science. http://www.iom.edu/frackingHIA.

Panko, J.M., et al. 2011. Natural Gas Extraction Using Hydraulic Fracturing: Creating a Sustainable Path Forward, ChemRisk. http://www.chemrisk.com/HF%20White%20Paper%20August%2017.pdf.

Reese, S. O., and Risser, D. W. 2010. Summary of Groundwater-Recharge Estimates for Pennsylvania. Water Resource Report 70. Pennsylvania Geological Survey. Harrisburg. http://www.dcnr.state.pa.us/topogeo/pub/water/pdfs/w070.pdf.

Swackhamer, D. L. 2012. Potential Impacts of Hydraulic Fracturing on Water Resources. University of Minnesota Workshop on the Health Impact Assessment of New Energy Sources: Shale Gas Extraction. National Academy of Science. April 30, 2012. http://www.iom.edu/frackingHIA.

Sumi, L. 2008. Shale Gas: Focus on the Marcellus Shale. Oil and Gas Accountability Project. Earthworks. http://www.earthworksaction.org/pubs/OGAPMarcellusShaleReport-6-12-08.pdf.

Vidic, R. 2010. Sustainable Water Management for Marcellus Shale Development. Dept. of Civil and Environmental Engineering, University of Pittsburgh. http://www.temple.edu/environment/NRDP_pics/shale/presentations_TUsummit/Vidic-Temple-2010.pdf.

Volz, Conrad. 2010. Marcellus Shale Gas Extraction; Public Health Impacts and Visualization of Environmental Threats. University of Pittsburgh Graduate School of Public Health. http://energy.wilkes.edu/PDFFiles/Issues/Volz_Jan2011_ProgressSummit.pdf

THE PUBLIC HEALTH IMPLICATIONS OF UNCONVENTIONAL NATURAL GAS DRILLING

Bernard D. Goldstein[1*], Jill Kriesky[1]

[1]Department of Environmental and Occupational Health, Graduate School of Public Health, University of Pittsburgh, Pittsburgh, PA
*Email: bdgold@pitt.edu

ABSTRACT

The opportunity to capitalize on improved horizontal hydraulic fracturing technology to extract natural gas from shale formations across the country has led to rapid development of the industry. Its pace, combined with seeming government and industry reluctance to approach the subject, has outstripped that of research on the potential health impacts of drilling-related activities on workers, nearby residents and surrounding communities. We briefly discuss the multiple pathways through which human health may be affected by shale gas drilling and review the existing research literature on health impacts they pose. We conclude with recommendations for increased air, water, soil, noise, and community monitoring; and for implementation of comprehensive epidemiological studies of drilling and production communities and workers.

KEY WORDS: Hydraulic fracturing, public health, air pollution, water pollution, community health

INTRODUCTION

The rapid increase in the technology to extract natural gas bound to shale layers has led to major changes in the overall picture of fossil fuel availability in the United States and potentially elsewhere. The use of horizontal drilling and hydraulic fracturing (HF), which we will describe as unconventional gas drilling (UGD), has greatly increased the access to shale gas deposits. Horizontal wells increased from a few hundred to more than 10,000 in Texas alone from 2004-2010 (EIA 2011). It is estimated that over 20,000 shale gas wells were drilled in the United States in the span of 10 years ending in 2010 (MIT 2011). Continued rapid growth is expected. In 2010, shale gas accounted for 23% of the natural gas supply and is projected to increase to 49% by 2035. (EIA 2012).

By 2008, the Marcellus Shale play under large parts of Pennsylvania, West Virginia, New York, and smaller parts of Ohio, Maryland, and Virginia, became the target of fast development. The Marcellus is the most expansive of shale gas plays in the US with estimates of 489 trillion cubic feet (TCF) of recoverable gas (Kargbo 2010).

Natural gas has been supported as a means of decreasing pollutant emissions, including carbon dioxide, by replacing other fossil fuels, particularly coal. However, the population density and proximity to the drinking water sources for major east coast cities of some of the Marcellus well sites, along with the growing number of reports of environmental and health impacts in southern and western locations has made shale gas drilling an increasingly high-profile and controversial activity. The release of the anti-shale gas drilling documentary Gasland in 2010, followed by comprehensive investigative reporting by the lay and scientific news media is making "fracking" a household word (Goldstein *et al.* 2012, Lustgarten and ProPublica 2011, Urbina 2011, Goldstein and Kriesky 2012, Subra 2009, Subra 2010).

More recently, community-based organizations have urged state and federal officials to enact regulations or moratoria based on concerns about alleged environmental and health impacts and changes in social infrastructures in communities near wells, compressors, and pipelines. In response, states including Pennsylvania and Maryland, and the federal government have assembled advisory commissions to weigh these opposing positions as they develop public policy related to the industry. Despite pronouncements by the respective governors and the President signaling the intent to address potential public health impacts in these panel studies, none of their commissions have included health professionals (Goldstein *et al.* 2012). Indeed, in Pennsylvania the recently enacted impact fee will be distributed among 17 different state and local agencies, subagencies and commissions, but not the Pennsylvania Department of Health which previously had been slated to develop registries and other approaches to health complaints.

As a result of the lack of relevant research, this overview of the potential health impacts of shale gas extraction approaches the issue indirectly. We assume that the potential pathways for human health impacts include contamination of air, water, and soil on or in close proximity to well pads, compressors and pipelines, as well as upstream and downstream events. Potential health consequences include acute injuries from events such as explosions and construction-related accidents at well sites and traffic accidents involving the thousands of trucks traveling to these sites carrying water, hydrofracking agents, and construction equipment. A range of social impacts, from noise that disrupts sleep and the ability to concentrate to the sudden influx of a mostly male transient workforce, represents a third avenue for impacts on human health. We briefly summarize the information related to each of these pathways to health impacts and make recommendations for future research. Understanding of the relationship between gas extraction and health is necessary for policy development that both nurtures the emerging, strategic industry and protects humans potentially affected by it.

HEALTH EFFECTS STUDIES

There is a dearth of formal, epidemiological protocols or health impact assessments in place or even contemplated in shale gas drilling communities. We briefly review a few of the studies, but emphasize that none of these is definitive in establishing a link between shale gas drilling and observed health impacts. They do provide some direction for the development of comprehensive studies that could identify the relationship between the inputs and outputs to drilling and human health. Witter *et al.* (2008) have reviewed the literature on the potential exposure-related health impacts of oil and gas development. Their subsequent Health Impact Assessment of Battlement Mesa, Colorado is the only comprehensive study of its kind in that it thoroughly evaluates the potential adverse consequences from the wide range of pollutant and other stressors (Witter *et al.* 2010). We have also gathered (but not yet published) survey data about perceived health impacts from individuals residing in close proximity to drilling sites has also been. Our findings are similar to those of Subra who has described complaints involving virtually every body organ system in residents of communities with UGD (Subra 2010). Although none of these studies have been published in a peer-reviewed journal, they do document that there are people who believe their health has been affected, and are consistent with the statements of citizens who have testified against shale gas drilling (Goldstein *et al*. 2012). Further, the findings serve as a basis for the needed thorough epidemiological evaluation that has yet to begin. Evidence that domestic and farm animals have been affected has been reported, with presumed implications to human health (Bamberger and Oswald 2012).

Shale gas development may also benefit the health of the community through improving the economic well-being of what are now depressed rural areas. More information about the longer term economic benefits and its potential health value are needed. The overall issue begs for a sustainability approach, such as the framework recently suggested by the National Research Council for EPA (NRC 2011). Balancing the economic, environmental and social/health impacts of UGD should be approached in a systematic way that considers longer term as well as shorter term impacts.

PATHWAYS OF EXPOSURE AND DIRECT CHEMICAL AND PHYSICAL EFFECTS RELATED TO UGD

We summarize many of the health issues addressed elsewhere in this paper in Table 1. The potential for adverse health impacts of UGD extends well beyond the drilling process. Upstream activities include providing the chemical and physical agents used at the drill site. For example, NIOSH is evaluating potential silicosis in miners in the Midwest who obtain the silica used as a proppant, as well as the exposure to silica of workers at the drill site. Downstream activities include the major issue of disposition of flowback and produce water, as well as issues related to pipelines and the eventual delivery of a gas product. For example, will there be more radon in natural gas delivered to major east coast cities due to less time for radioactive decay when the natural gas source is nearby rather than more than a thousand miles away in the south? (Resnikoff 2012)

Table 1. Examples of Potential Pathways for Human Health Impacts Related to Unconventional Shale Gas Drilling

Safety Issues	Worker and Community	Construction/drilling incidents
		Fires and explosions
		Traffic Incidents
Air Pollution	Worker exposure	HF chemicals
		Silica
		Diesel exhaust
		Drilling compounds
	Community exposure	HF chemicals
		Hydrocarbons/air toxics
		Nitrogen oxides
		Diesel

	Regional exposure	Ozone
	Global climate change	Methane
Water Pollution	*Drill site*	Release of HF chemicals deep underground (unproven)
		Release of HF chemicals at or near surface
		Diesel spills
		Onsite release of flowback and product water
	Offsite flowback and product water	Impoundment and contaminant releases
		Treatment and recycling issues
Agents of Concern	*Hydrofracturing agents*	Evolving list of chemicals
	Hydrocarbons	Methane
		Benzene
		Other volatile hydrocarbons
	Naturally occurring agents released to the surface	Brine salts
		Bromine
		Barium
		Arsenic
		Minerals
		Radioactive agents
	Noise	Stress effects
Surprises	*Unforeseen exposure and effects*	Chemical mixtures
		Chemical reactants
		TENORM*
		Earthquakes
		Ecosystem disruption leading to human health effects

*TENORM: Technically enhanced naturally occurring radioactive materials

Colborn *et al.* (2011) have investigated the potential contamination of air, water, and soil resulting from the fracking process by reviewing primarily the Material Safety Data Sheets (MSDS) of chemicals used in shale gas extraction and relating these to databases on health impacts of these agents. They report on the potential for effects on various organ systems and particularly raise concerns about endocrine disruptive agents, which are also of concern to Finkel and Law (2011). However, there are no data on the exposure dose of these agents on which one can base a risk assessment.

Exposure assessment is a central approach to evaluating the potential adverse health consequences of chemical and physical agents released as a result of UGD. A complete exposure pathway can be demonstrated when the following five elements are present: sources of contaminants; environmental media containing the contaminants; points of exposure; routes of exposure; and a receptor population. We are unaware of any comprehensive approach to determine if a complete exposure pathway exists. Studies evaluating exposure pathways that have demonstrated potential health risk include those of McKenzie *et al.* (2012) who used standard EPA risk-based approaches to estimate the risks for air emissions. They reported elevated subchronic non-cancer risks, including effects on the nervous system, to community residents living near the well site and an additional cancer risk of 10 in 1 million lifetime, primarily due to benzene. Highest risks occurred due to air emissions during the hydraulic fracturing completion phase. Research on water pollution has included observation of higher levels of methane in well water ascribed to shale gas drilling (Osborn *et al.* 2011).

In this review we do not use the standard approach in toxicology of addressing one-by-one the chemical and physical agents of concern. While valuable, the shortcomings in using this approach to the issue of the potential toxicological health effects of UGD include the fact that the HF agents are changing rapidly as the process evolves, that they are too often secret, and that local geology impacts the agents that are used and both the hydrocarbons and natural agents that come to the surface. Most importantly, however, is that a one-by-one approach to these various agents loses sight of what we believe are a highly likely cause of significant direct toxicological impact - that of the effect of the mixtures of agents. Unfortunately, the USEPA will not be considering mixtures in its first round of evaluation which focuses on HF agents and which is expected to be completed in 2014. Applying newer molecular toxicological advances to understanding the threat of mixtures in relation to UGD should be a high priority.

WORKER HEALTH AND SAFETY ISSUES AND THEIR COMMUNITY IMPLICATIONS

In addition to having safety issues similar to other large construction sites, the hydrocarbons present at drilling sites are sources of potential fire or explosion. Additional workplace health and safety issues are related to HF chemical and physical agents, the drilling process, and to the increasingly high pressures used in HF. Previous studies showed substantial and increasing worker risks associated with the drilling industry (CDC 2008). Unfortunately, the penchant to use temporary employees and out-of-area workers complicates obtaining the necessary information on which to base workplace health and safety initiatives.

Some of the dangers to worker health are those that also put community members at risk. These include fires and explosions as well as truck safety issues. Further, workers often are community members.

AIR POLLUTION

Associated with shale gas drilling activities are a number of different pathways by which pollutants may be released to the air and inhaled by workers or community members. Upstream activities include processes involved with the hydrofracturing agents before they reach the site, such as their manufacturing and trucking. In addition to silica, inhalable agents at the site include the HF components; diesel exhaust, which is now considered to be a known human carcinogen; and the hydrocarbon components of natural gas. Particularly in wet gas areas, such as Southwestern Pennsylvania, benzene will be among the higher molecular weight components of natural gas. Benzene is a known cause of human leukemia, and is also a common part of our petrochemical era. It is present in much higher levels in crude petroleum and gasoline than it is likely to be present in shale gas. Benzene in air or water will inevitably be of concern to community members as will other air toxics that are released from the drilling procedures and from downstream activities.

Air pollution as a result of UGD may have regional or global consequences. Ozone is formed by the action of sunlight on oxides of nitrogen (NOx) and hydrocarbons, often occurring literally hundreds of kilometers downwind from the original sources. It is a particularly potent oxidant causing lung damage and respiratory problems such as childhood asthma attacks. Regulatory requirements for ozone are of particular concern for areas such as Texas and the Northeast where exceedances of the ozone standard have been common in the recent past. As the ozone standard is likely to become more stringent due to increasing evidence of low dose toxicity, it is at least possible that the additional hydrocarbons and NOx released from the many individual well sites will cause an exceedance of the new standard. Regulating this aggregate risk of perhaps thousands of sites within a region is challenging as current regulations are focused on individual large sources, such as refineries.

Methane, the principal component of natural gas, is a very potent greenhouse gas. To the extent that methane is released as part of the drilling or distribution process, it counteracts the potential benefit of natural gas in decreasing carbon dioxide emissions as compared to burning other fossil fuels.

WATER POLLUTION

Shale gas drilling is mainly occurring in rural areas in which the population is served by small wells tapped into local groundwater resources. Concern about hydrofracking chemicals polluting groundwater resources is not surprising. Confusion about this issue is caused by varying definitions of HF. If restricted to successful injection of HF agents into shale layers that are deep underground and far below groundwater sources, there is no proof that HF agents can migrate to groundwater sources. However, there is no question that HF agents have contaminated well water as a result of incidents at or near the surface, e.g., blown casings, ruptured drums, and there is a controversial issue of leakage into groundwater from relatively shallow shale gas wells drilled in Pavilion, Wyoming (EPA 2011).

In our estimation, the greater long term concern for water pollution is the issue of what contaminants are contained in, and what to do with, the flowback and produce waters. Flowback water consists of the relatively immediate return of perhaps 20-60% of the upwards of 5 million gallons used in the hydrofracturing process. This flowback water contains not only the return of the agents injected underground, but also the hydrocarbons that are the targets of the drilling process and the many chemical and physical agents that are naturally present underground. It is these naturally present agents, and reactants formed among various injected and naturally present agents, that we believe present the greatest health threat. Produce waters are the much smaller volumes produced daily during the lifetime of the well but which in aggregate in a high drilling region also present challenges to safe disposal Table 1 contains a partial listing of the chemical and physical agents of concern. Contributing to our lack of knowledge about flowback and produce water contaminants are state laws which absolve the drilling industry from revealing information about natural agents or the results of chemical reactions (Goldstein and Kriesky 2012).

The flowback and produce waters deserve particular attention because they are what comes back up to the surface and must be contained and disposed of appropriately. A particular challenge is the high brine content. Some disposal approaches have already led to problems. For example, after problems were noted, Pennsylvania instituted a voluntary moratorium on using POTWs for flowback waters. However, illustrative of the surprise element, trucking the water for disposal in deep wells in Ohio, appears to have led to earthquakes.

When hydrofracturing is defined as the successful release of HF agents perhaps one mile underground, there is little evidence that this will lead to contamination of groundwater. But the public's concern is whether their groundwater will become contaminated from any part of the entire UGD process - from the time the drill pad is leveled until perhaps decades later when the gas runs out. Producing confusion and anger in the public is the current situation in which they are told that there is no proof that HF ever leads to groundwater contamination, yet they read about communities whose well water has been contaminated by HF agents. The fact that such contamination has occurred from surface or near surface carelessness is immaterial to the basic concern of the public of whether groundwater will be contaminated by drilling activities.

SURPRISES

In view of the complex evolving nature of shale gas drilling with new chemicals and the ever deeper penetration of the earth, surprises are inevitable. Two so far have been the high bromine content of the wastewater which could lead to brominated compounds of potentially high toxicity in drinking water, and the earthquakes apparently due to deep injection of flowback water.

Other areas that fit under the surprise category are unforeseen effects of mixtures of agents. Also of concern is the potential for problems related to radioactivity within the shale layers that could be displaced into homes near or far.

PSYCHOSOCIAL DISRUPTION AS A CAUSE OF HEALTH EFFECTS

There is no question that the public's concern about health effects is a substantial reason for opposition to UGD (Goldstein *et al.* 2012). Direct neurotoxic effects, such as chemical exposures that cause headache and nausea, can be difficult to differentiate from those due to the mental health effects of chemical exposures, a problem that has been noted after oil spills such as that of the Deepwater Horizon (Goldstein *et al.* 2011). Stress can also be a substantial cause of adverse health effects.

Evidence of psychosocial disruption in shale gas communities has been obtained by a number of researchers. Researchers identify a variety of sources and measures of such disruption ranging from increased traffic, caseloads for social service agencies and housing shortages to increased crime all resulting from the influx of gas workers and their families. The loss of the simple rural lifestyle valued by many local residents and divisions between citizens whose economic prospects improve with drilling -- either through employment or mineral rights leasing -- and those whose do not likewise increase stress which can impact health (Brasier *et al.* 2011, Anderson and Theodori 2009, Blevins *et al.* 2005, Jacquet 2009). Perceived or actual economic losses in agriculture and tourism can similarly disrupt communities as drilling commences (Brasier *et al.* 2011, Perry 2011).

From the vantage point of public health, these health complaints cannot be discounted just because there is no direct evidence of exposure to chemical or physical agents resulting from UGD. Health is defined by the World Health Organization as "a state of complete physical, mental, and social well-being and not merely the absence of disease or infirmary" (Table 2) (WHO 1948). Moreover, the WHO definition of health encompasses recognition of the social determinants of health (Table 2, WHO 1948, CSDH 2008). Accordingly, under the standard approach to public health, there are people who are not healthy due to UGD—and this is true whether the cause is their outrage at the changes in their communities (Sandman, 1987), their mistrust of government or industry, their feelings of helplessness, or to the direct effects of chemical or physical agents.

Table 2. Public Health

WHO Definition of Health (1948)	WHO Commission on Social Determinants of Health (CSDH 2008)
A state of complete physical, mental, and social well-being and not merely the absence of disease or infirmary.	…the circumstance in which people are born, grow up, live, work and age, and the systems put in place to deal with illness. These circumstances are in turn shaped by a wider set of forces: economics, social policies and politics.

A large literature exists documenting the importance of trust to public's response to a potential health threat(Sandman 1987, Fischhoff *et al.* 1984, Peters *et al.* 1997). Industry and government need to work hard to demonstrate that they are worthy of trust including incorporating the public in their planning; acting openly and transparently; and providing firm evidence of their concern for public well-being and respect for community values. Secrecy is particularly problematic when public trust is involved. While there has been some movement toward divulging HF agents employed at a specific drilling site, the persistent use of "confidential business information" as an excuse for secrecy, and the wall that industry is building around releasing information about the natural agents that are "incidentally" pulled up from underground, or about the result of reactions among the various agents, continues to be detrimental toward establishing trust (Goldstein and Kriesky 2012, PPUC 2012). Confusing messages, such as whether UGD is a new technological advance that now permits extracting gas from tight shale layers deep underground; or has been around for decades so there is no need for concern, are also not helpful in obtaining public trust. Also problematic is a triumphalism that is exemplified by the head of an industry organization recently being quoted as stating that "Every Pennsylvanian benefits" from shale gas drilling. Those made homeless by higher rental costs due to the influx of drill site workers are not Pennsylvanians who benefit (Williamson and Kolb 2011).

To maximize the benefits of UGD the legitimate concern about potential health implications must be addressed. To do so requires independent assessment of public health impacts. Such an assessment should include a broad approach to public health endpoints, and should involve the community in its research. Informing the community should include health education programs for residents as well as for workers and health care professionals.

To evaluate potential adverse health effects from hydraulic fracturing and extraction of natural gas will require systematic and thorough health impact assessments including epidemiological studies based upon thorough understanding of exposure pathways and of body burdens. Variation across shale plays in the geology, geography and weather conditions, fracking agents used, and mixtures returned to the surface requires consideration of local conditions. The potential health impacts of the shale gas industry consists not only of activities on the drilling site but also include the broad range of upstream and downstream activities and the psychosocial impacts on the community.

REFERENCES

Anderson, B.J.; Theodori, G.L. (2009) Local Leaders' Perceptions of Energy Development in the Barnett Shale. *Southern Rural Sociology*. **24**, 113.

Bamberger, M.; Oswald, R.E. (2012) Impacts of Gas Drilling on Human and Animal Health. *New Solut*. **22**, 51-77.

Blevins, A.; Jensen, K.; Coburn, M.; Utz, S. (2004) Social and Economic Impact Assessment of Sublette and Sweetwater Counties. Wyoming: University of Wyoming. Available from: http://www.sublettewyo.com/DocumentCenter/Home/View/368.

Brasier, K.J.; Filteau, M.R.; Jacquet, J.; Stedman, R.C.; Kelsey, T.W.; Goetz, S.J. (2011) Residents' Perceptions of Community and Environmental Impacts From Development of Natural Gas in the Marcellus Shale: A Comparison of Pennsylvania and New York Cases. *Journal of Rural Social Sciences*. **26**, 32-61.

Centers for Disease Control and Prevention (CDC). (2008) *Fatalities Among Oil and Gas Extraction Workers--United States, 2003-2006*. MMWR.Morbidity and Mortality Weekly Report. **57**(16), 429-431.

Colborn, T.; Kwiatkowskia, C.; Schultza, K.; Bachrana, M. (2011) Natural Gas Operations from a Public Health Perspective. *Hum Ecol Risk Assess*. **17**, 1039-1056.

Commission on Social Determinants of Health (CSDH) (2008) *Closing the Gap in a Generation: Health Equity Through Action on the Social Determinants of Health. Final report of the Commission on Social Determinants of Health*. Geneva, World Health Organization. Available from: http://www.who.int/social_determinants/thecommission/finalreport/en/index.html.

Finkel, M.L.; Law, A. (2011) The Rush to Drill for Natural Gas: A Public Health Cautionary Tale. *Am J Public Health*. **101**, 784-785.

Fischhoff, B., Hope, C., & Watson, S. R. (1984) *Defining Risk*. Policy Sciences. **17**(2), 30-139.

Goldstein, B.D.; Kriesky, J. (2012) The Pennsylvania Gas Law Fails to Protect Public Health. *Pittsburgh Post-Gazette.* Mar 11, 2012. Available from: http://old.post-gazette.com/pg/12071/1215612-109.stm.

Goldstein, B.D.; Kriesky, J.; Pavliakova, B. (2012) Missing from the Table: Role of the Environmental Public Health Community in Governmental Advisory Commissions Related to Marcellus Shale Drilling. *Environ Health Perspect.* **120**, 483-486.

Goldstein, B.D.; Osofsky, H.J.; Lichtveld, M.Y. (2011) The Gulf Oil Spill. *N Engl J Med.* **364**, 1334-1348.

Jacquet, J. (2009) Energy Boomtowns and Natural Gas: Implications for Marcellus Shale Local Governments and Rural Communities. State College, PA: Northeast Regional Center for Rural Development; 43. Available from: http://nercrd.psu.edu/Publications/rdppapers/rdp43.pdf.

Kargbo, D.M.; Wilhelm, R.G.; Campbell, D.J. (2010) Natural Gas Plays in the Marcellus Shale: Challenges and Potential Opportunities. *Environ Sci Technol.* **44**, 5679-5684.

Lustgarten, A.; ProPublica. (2011) Climate Benefits of Natural Gas may be Overstated. http://www.scientificamerican.com/article.cfm?id=climate-benefits-natural-gas-overstated. (accessed Jul 9, 2012).

McKenzie, L.M.; Witter, R.Z.; Newman, L.S.; Adgate, J.L. (2012) Human Health Risk Assessment of Air Emissions From Development of Unconventional Natural Gas Resources. *Sci Total Environ.* **424**, 79-87.

Massachusetts Institute of Technology Energy Initiative (MIT). (2011) *The Future of Natural Gas* (Interdisciplinary study). Massachusetts: Massachusetts Institute of Technology.

National Research Council (NRC). (2011) *Sustainability and the U.S. EPA.* Washington, D.C.: National Academies Press.

Osborn, S.G.; Vengosh, A.; Warner, N.R.; Jackson, R.B. (2011) Methane Contamination of Drinking Water Accompanying Gas-Well Drilling and Hydraulic Fracturing. *Proc Natl Acad Sci U S A.* **108**, 8172-8176.

Pennsylvania Public Utility Commission (PPUC). (2012) *Act 13 (Impact Fee).* http://www.puc.state.pa.us/naturalgas/naturalgas_marcellus_Shale.aspx. (accessed Jul 9, 2012).

Perry, S. L. (2011) *Energy Consequences and Conflicts Across the Global Contryside: North American Agricultural Perspectives.* Forum on Public Policy. **2011**(2), 11 July 2012. Available from: http://forumonpublicpolicy.com/vol2011.no2/archivevol2011.no2/perry.pdf

Peters, R. G., Covello, V. T., & McCallum, D. B. (1997) *The Determinants of Trust and Credibility in Environmental Risk Communication: An empirical study*. Risk Analysis : An Official Publication of the Society for Risk Analysis. **17**(1), 43-54.

Resnikoff, M. (2012) *Radon in Natural Gas From Marcellus Shale*. Radioactive Waste Management Associates.

Sandman, P. (1987) Risk Communication: Facing Public Outrage. *US Environmental Protection Agency Journal*. **12**, 21-22.

Subra, W. (2010) Community Health Survey Results, Pavillion, Wyoming Residents. *Earthworks' Oil and Gas Accountability Project*.

Subra, W. (2009) Health Survey Results of Current and Former DISH/Clark,Texas Residents. *Earthworks' Oil and Gas Accountability Project, New Iberia, LA*.

Urbina, I. (2011-12) Drilling Down series. *New York Times*. Available from: http://www.nytimes.com/interactive/us/DRILLING_DOWN_SERIES.html.

U.S. Energy Information Administration (EIA). (2012) *Annual Energy Outlook 2012 With Projections to 2035* No. DOE/EIA-0383(2012). Washington, DC: US Department of Energy.

U.S. Energy Information Administration (EIA). (2011) *Technology Drives Natural Gas Production Growth From Shale Gas Formations*. http://www.eia.gov/todayinenergy/detail.cfm?id=2170 (accessed Jul 5, 2012).

U.S. Environmental Protection Agency (EPA). (2011) *Draft: Investigation of Ground Water Contamination Near Pavillion, Wyoming*. Ada, Oklahoma: U.S. EPA; EPA 600/R-00/000. Available from: http://www.epa.gov/region8/superfund/wy/pavillion/index.html.

Williamson, J.; Kolb, B. (2011) Marcellus Natural Gas Development's Effect on Housing in Pennsylvania. Williamsport, PA: Center for the Study of Community and the Economy.

Witter, R.; McKenzie, L.; Towle, M.; Sinson, K.; Scott, K.; Newman, L.; Adgate, J. (2010) Health Impact Assessment for Battlement Mesa, Garfield County Colorado. *Colorado School of Public Health*, University of Colorado Denver, Aurora, CO.

Witter, R.; Stinson, K.; Sackett, H.; Putter, S.; Kinney, G.; Teitelbaum, D.; Newman, L. (2008) Potential Exposure-Related Human Health Effects of Oil and Gas Development: A White Paper. http://docs.nrdc.org/health/files/hea_08091702a.pdf. (accessed Dec 14, 2011).

World Health Organization (WHO). (1948) Preamble to the Constitution of the World Health Organization as Adopted by the International Health Conference. New York: World Health Organization. http://www.who.int/about/definition/en/print.html. (accessed Jul 5, 2012).

RADIOACTIVITY IN MARCELLUS SHALE
CHALLENGE FOR REGULATORS AND WATER TREATMENT PLANTS

Marvin Resnikoff[a*]

[a]Radioactive Waste Management Associates, Bellows Falls, VT 05101, USA

* Email: radwaste@rwma.com

ABSTRACT

Studies by the U.S. Geological Survey and gamma logs of drillers show radium concentrations up to 32 times background concentrations in the Marcellus shale formation. Brought to the surface in rock cuttings, drilling fluids, flowback water and brine, radium can enter the environment in several forms. A fraction will be reinserted into deep disposal wells, or go to water treatment plants. Because of its high salinity, a portion will be spread on highways in the winter. The rock cuttings will go to solid waste landfills. Over the production cycle, a portion of Marcellus radium plates out on pipes. And an inert radioactive gas, radon, will enter homes when natural gas is used for heating and cooking.

This paper examines the fate and transport of radionuclides brought to the surface, and the environmental impact in the environment. Regulators have the task of developing regulations that protect the health and safety of the population. The water treatment industry must develop processes for separating radium from large volumes of waste waters so that surface streams meet regulatory limits. This paper explores these different waste streams and potential resolutions of these differing problems.

KEYWORDS: Marcellus shale, natural gas, radioactivity, fate and transport, regulation

INTRODUCTION

Geologists consider the Marcellus shale formation to be relatively highly radioactive and regionally extensive. Radioactivity in the Marcellus shale results from the high content of naturally occurring radioactive uranium and thorium, including their decay products, and potassium elements in the rock. In New York State the formation ranges from 25 to over 100 feet in thickness and the depth of the base varies from an outcrop to 1000-foot depth by Syracuse to 4000-foot depth by the border with Pennsylvania (Hill, 2994).

The areal extent of Marcellus shale is shown in Figure 1. As seen, the formation underlies the States of New York, Pennsylvania, eastern Ohio, West Virginia, Virginia, all the way into Alabama.

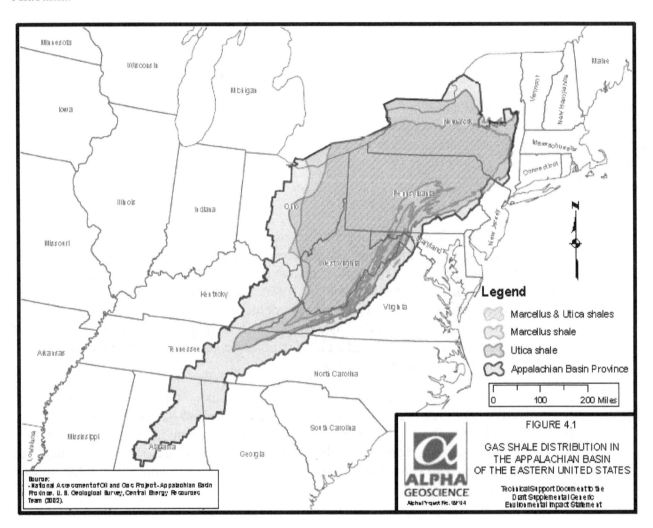

Figure 1. Areal Extent of Marcellus Shale (NYSERDA, 2009)

The depth to the top of the Marcellus formation in New York State is shown in Figure 2. As seen, the formation is deeper as one goes south.

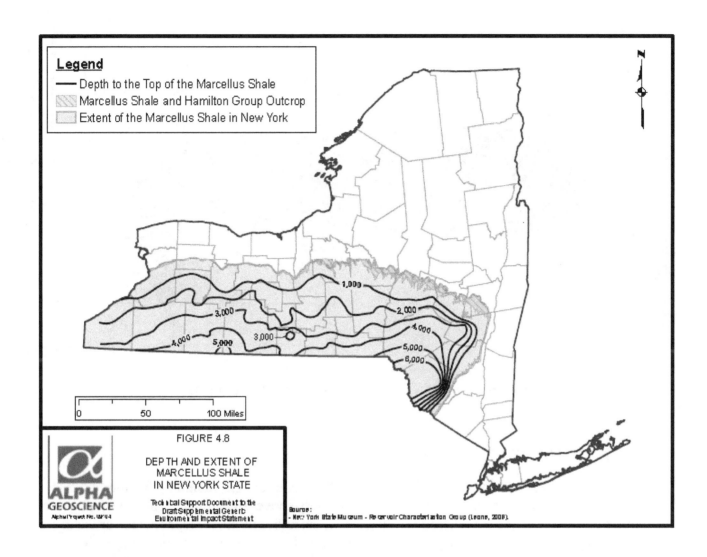

Figure 2. Depth of Marcellus Shale in New York State (NYSERDA, 2009)

In addition to the depth of the Marcellus formation, the thickness varies as well. In New York State the Marcellus shale thickness increases from less than 25 feet in the western section of the State, Chautauqua County, to 300 feet in the southeastern section of the State, as seen in Figure 3. The dots represent exploratory gas wells.

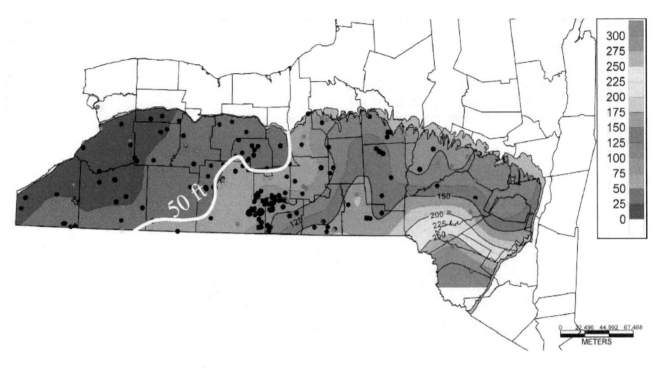

Figure 3. Marcellus shale thickness in New York (NYSERDA, 2009)

RADIOACTIVITY IN MARCELLUS SHALE

Most disinterested scientists agree that the Marcellus shale contains uranium-238, thorium-232 and their decay products at concentrations significantly above background. The issue is: how much radioactivity above background? In general, the radioactivity throughout the vertical depth of rock cuttings appears to be equal to or less than 10 picocuries per gram, including K-40. However, at certain depths in each well, the activity is significantly higher. The gas industry knows it has reached the Marcellus shale horizon when cores show high gamma ray and high total organic carbon content (TOC). Consider the log from Beaver Meadows (Figure 4). This well is located near Norwich in Chenango County in upstate New York.

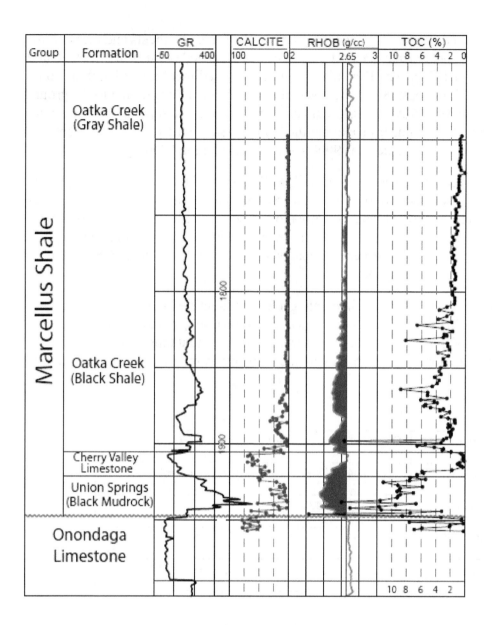

Figure 4. Beaver Meadows Core (AAPG, 2010)

As one goes from the shale layer (Union Springs) to the limestone layer (Onondaga Limestone), one sees a major drop in gamma ray and total organic content. Without going into a discussion of GAPI units, 16.5 GAPI represents 1 pCi/g radium, and 400 GAPI is 24.2 pCi/g radium. As you can see, the GAPI is much greater than 400 units, and this also corresponds to very high TOC. These radium concentrations are far higher than background radium concentrations in New York State (Myrik, 1981), which are 0.85 pCi/g. Other gamma ray logs wells down to the Marcellus shale horizon in New York State such as the Shiavone 2, WGI11 and Bergstrasser wells show similar results.

The logs mentioned above correspond to a geochemical study of trace elements and uranium in the Devonian shale of the Appalachian Basin (Leventhal, 1981) carried out in 1981 by the United States Geological Survey (USGS). The Devonian layer refers to sediment formed 350 million years ago from mud in shallow seas. Since the layers do not form in a line parallel to the ground surface, the depth at which Marcellus is found can vary from surface outcroppings to as deep as 7,000 feet or more below the ground surface along the Pennsylvania border in the Delaware River valley (NYDEC, 2011), and as deep as 9000 feet in Pennsylvania (Geology, 2008).

The USGS study analyzed seventeen cores from wells in Pennsylvania, New York, Ohio, West Virginia, Kentucky, Tennessee, and Illinois. The researchers collected a variety of geochemical data to be used for resource assessment and identification of possible environmental problems. Rather than direct gamma spectroscopy, uranium was measured in each core with a more appropriate and precise method, delayed-neutron analysis. I use data from the USGS since it is a reputable and objective government agency.

Although the cores varied in thickness and in depth, geologists identified the Marcellus Shale stratum in several cores using data on the organic matter (carbon), sulfur, and uranium content of the samples. Table 1 below summarizes the results from four cores that tapped into the radioactive Marcellus formation. The depths at which the layer was found as well as the uranium measurements are presented.

Location of the Core	Depth of Sample (feet)	Uranium Content (ppm)
Allegheny Cty, PA	7342 – 7465	8.9 – 67.7
Tomkins Cty, NY	1380 – 1420	25 – 53
Livingston Cty, NY	543 – 576	16.6 – 83.7
Knox Cty, OH	1027 – 1127	32.5 – 41.1

Table 1. Uranium Content and Depth of Marcellus Shale in Four Cores

The four cores were taken from different geographical locations, but the characteristics of the identified Marcellus shale layer, specifically the high uranium and carbon content, are consistent.

To compare the uranium content in parts per million (weight) to radioactive concentration in picocuries per gram, we use the correspondence (HPS, 2007):

$$2.97 \text{ ppm} = 1 \text{ pCi/g U-238}$$

Using this relationship, the U-238 ranges up to 28 pCi/g, or 32 times background for radium-226, assuming U-238 and Ra-226 are in secular equilibrium, as they are in the Marcellus Shale formation. That is, the USGS measurements and the GAPI logs are consistent. This is our starting point for the concentrations of Ra-226 in the natural Marcellus Shale formation. The radium itself is found in the pore water of the Marcellus Shale formation, since radium, under the temperature, pressure and chemical conditions, is preferentially dissolved in the pore water.

FATE AND TRANSPORT OF RADIOACTIVE MATERIALS

If higher levels of radioactivity are present in the Marcellus shale formation, as shown by gamma logs, USGS measurements, and statements by DEC, where will this radioactivity appear in the aboveground environment and how will NYDEC and PADEP regulate this radioactive waste?

To discuss where uranium, radium and other radionclides enter the environment, we consider two components of natural gas development: well development and production.

Well Development

We are concerned about the horizontal, not the vertical, component of natural gas wells. As seen in Figure 5, fracking consists of pumping a fluid, proppants and chemicals under high pressure to create fractures along the horizontal region. I'm not going to discuss other chemicals that may contaminate properties, such as arsenic, mercury and hydrocarbons. The fracture zone may extend up to 200 foot radius from the horizontal bore. The bore itself may extend horizontally up to a mile. One 4,000-foot lateral wellbore undergoing hydraulic fracturing may require between 2.4 million and 7.8 million gallons of water (NYDEC, 2009). During well development, rock cuttings are produced. These are separated from the drilling fluid by shakers, shown in Figure 6.

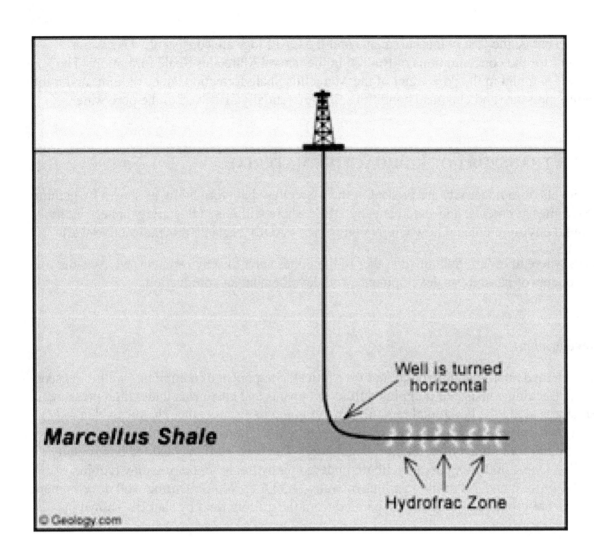

Figure 5. Fracking

Figure 6. Rock cutting shakers

Rock Cuttings

The drilling fluid itself may be partially recycled and the rock cuttings have been going to solid waste landfills. Radium is primarily located in the water and not in the shale rock itself. But solid waste that goes to a solid waste landfill can be, according to NYS regulations, up to 20% water. An industry lab measured rock cuttings with a gamma detector; the rock cuttings were shown to be slightly radioactive. On the other hand, NYDEC, in their 2009 Draft Supplemental GEIS, reported radioactivity for two wells in Lebanon and Bath, NY, total radioactivity concentrations of 25.4 +/- 4.6 and 29.2 +/- 4.3 pCi/g, respectively, which is consistent with what we expect from the gamma logs and USGS findings. The EPA limits for radium in surface soil (EPA, 2011) are 5 pCi/g in the top 15 cm, and 15 pCi/g below 15 cm. In my opinion, these rock cuttings, above EPA limits, should be disposed of in a licensed facility for handling radioactive materials, such as Energy Solutions in Clive, Utah. DEC and the Ohio EPA believe they have no jurisdiction over NORM, but if drilling fluid is recycled, the NORM is technologically enhanced, or TENORM. I'm not a lawyer, but it appears to me that these agencies do have jurisdiction over TENORM, if recycled drilling fluids are intermixed with rock cuttings. Some drilling materials disposed of in landfills in Ohio have set off entry radiation detectors.

Flowback Water

After a well is drilled, part of the hydraulic fracturing fluid flows back to the wellhead. This flowback water constitutes about 10 – 40% of the original pumped volume. (NYTimes, 2011) Since Marcellus shale is of marine origin, it contains high levels of salt and NORM which is dissolved in the fluid (Cornell, 2010). DEC found 13 samples of flowback water from vertical Marcellus shale wells in Schuyler, Chemung and Chenango counties to contain radioactive concentrations as high as 267 times the limit for discharge into the environment and thousands of times the limit for drinking water. (Davies, 2009) Flowback water from horizontal drilling will be much more radioactive. If flowback water is fed to a water treatment plant, the radium needs to be removed, as we'll discuss shortly.

Well Production

Once a natural gas well is producing, salt water or brine is brought up with the natural gas, and is separated out at the water gas separator. DEC has sampled brine for radium-226 and radium-228. The results are shown in Table 2 below. Collecting unpublished data from PADEP, USGS found radium concentrations in produced water (brine) ranging up to 20,000 pCi/L (USGS, 2011).

Well	Town	County	Ra-226	Ra-228
Maxwell 1C	Caton	Steuben	7885	234
Frost 2	Orange	Schuyler	2647	782
Carpenter 1	Troupsburg	Steuben	5352	138
Webster T1	Orange	Schuyler	16030	912
Calabro T1	Orange	Schuyler	13510	929
Schiavone 2	Reading	Schuyler	15140	957
WGI 11	Dix	Schuyler	10160	1252

Table 2. Radioactivity in Brine Water (pCi/L)

The radium concentrations in brine are very high. While no one is drinking brine, just to compare these concentrations with drinking water standards, 5 pCi/L combined radium-226 and 228. That is, the radium concentrations in brine range up to 3000 times safe drinking water limits. Under the proposed DEC regulations (NYDEC, 2010), a drill applicant must have a plan to deal with flowback water and brine, and that is where water treatment specialists enter the picture. A practice, which has since been suspended, is to spread brine, as a de-icing solution, on highways in New York State.

Just to get a handle on the magnitude of the problem. If hydraulic fracturing were approved, DEC estimates 1600 drilling applications per year; each horizontal well will produce between 2.4 million and 7.8 million gallons of flowback water. So we estimate 3.8 to 12.5 billion gallons of contaminated water per year that has to be treated. This is a small volume compared to the water usage in New York State, but this type of waste poses additional problems.

In Pennsylvania, much of the 1.3 billion gallons of radioactively-contaminated water produced between 2007 and 2010 was sent to sewage treatment plants. Most sewage plants in the State are not required to monitor for radioactivity, so there are little data. According to sewage plant operators interviewed by the Times (NYTimes, 2011), these plants are not equipped to remove radioactive material.

As a potential method of treating flowback water and brine, one can investigate the method of treating uranium mine waste water, which has high concentrations of radium. Here radium is precipitated with barium sulfate $Ba(Ra)SO_4$ by adding barium chloride to uranium mill waste water. Radium is then a particulate that can be filtered. Water treatment plants may then meet SPDES permit limits, and the filters will have to be sent to a licensed landfill. Other methods include cation exchange softening, lime softening, and sorption onto MnO_2. A physical process being proposed by Atela, Inc and Casella Waste Systems is distillation, low pressure evaporation. The sludge would then have to be disposed of in a licensed landfill. Volatiles from this process would have to be captured. It is not clear whether municipal water treatment plants can handle the magnitude of the radioactive waste problem posed by flowback water and brine, so that will be a challenge to water treatment professionals, such as you.

Radium scale buildup in gas equipment.
During production, radium dissolved in water, is brought to the surface. Scale, radium sulfate, plates out on production pipe surfaces. Scale also appears in water/gas separators, feeder lines and condensate tanks. As DEC states, a high concentration of scale will result in an elevated radiation exposure level at the pipe exterior surface (NYDEC, 2011). The NYS Department of Health (DOH) proposes a radioactive materials license when exposure levels exceed 50 microR/hr ($\mu R/h$) (NYDEC, 2012b). Our calculations show that the radium concentrations for an external dose rate of 50 $\mu R/h$ are far higher than the EPA limit of 5 pCi/g. We have found that exposed workers have an increased risk of developing cancer. Workers at, and residents near, pipeyards that clean pipe scale will have an additional risk from inhaling radioactive dust. Based on our experience, the proposed DOH regulations are too lax, as we discuss below.
For DEC and DOH to grasp the magnitude of the problem, we provide one example. At one natural gas well in Texas, 388 pipe joints (30' long) were pulled after 5 years service. Exposure levels exceeded 50 $\mu R/h$ in 55% of the 30 foot joints (max, 150 $\mu R/h$) 38% were < 50 $\mu R/h$ and 7% were free of NORM. Similarly, hundreds of pipes at each gas well in New York and Pennsylvania will be contaminated with radium scale. If thousands of gas wells are drilled in New York and Pennsylvania, how will NYDEC and PADEP have the resources to regulate the industry and track these contaminated pipes? In our experience, oil and gas producing pipes with high external exposure levels have been "donated" to city governments for playgrounds in Texas, or to farmers for use in animal corrals in Texas and Kentucky. They have been cut up with oxyacetylene torches and welded to fit their use. In the process, children and farmers have been directly exposed to gamma, and inhaled radium. Workers at pipeyards that cleaned pipes have inhaled radium-contaminated dust and have developed cancer.

55

We find that direct gamma exposure levels of 50 μR/h are much too high. In order to determine the concentrations of Ra-226 and Ra-228 that correspond to a dose rate of 50 μR/hr, we employed the program MicroShield (Grove, 2008), to estimate dose rates due to a specific external radiation source.

As inputs to MicroShield, we assumed an outer pipe diameter of 4 inches (10.16 cm), a scale thickness of 0.2 cm, and a pipe wall thickness of 0.91 cm, as suggested by the US EPA EPA, 1993b). We could also have used another standard pipe diameter, 2 7/8 inch. We assumed that each contaminated pipe is 30 feet long, and that radiation measurements had been taken at the center of the pipe, on contact with the outer pipe wall. From MicroShield, for a pipe with external gamma of 50 μR/h, we obtain a Ra-226 concentration in scale of 1,313.5 pCi/g, and a Ra-228 concentration in scale of 437.8 pCi/g that correspond to a dose rate of 50 μR/h. We assumed a 3 to 1 ratio of Ra-226 to Ra-228.

As NYDEC and PADEP are well aware, the EPA cleanup standard for total radium on soil is 5 pCi/g for the first 15 cm depth and 15 pCi/g 15 cm or more below the surface. Pipes, with external radiation 50 μR/h greatly exceed this standard. If these pipes, with NORM <50 μR/h are released for general use, they will be cut up and welded and the scale will be accessible. Ground contamination could easily exceed 5 pCi/g or 15 pCi/g. Though the 50 μR/h limit has been instituted in several States, in my opinion the NYDOH-suggested dose rate of 50 μR/h is much too high.

While DEC and DOH intend to regulate joints with external radiation > 50 μR/h, in practical terms, it is not clear what the State agencies envisage will become of these radioactive pipes (and separators, feed lines and condensate tanks). Once the oil and gas industry begins production in New York State, and sooner in Pennsylvania, this will be a major unresolved problem down the road.

Radon in Natural Gas

Yet another significant public health hazard associated with drilling for natural gas in the Marcellus Shale formation that should be seriously investigated by NYDEC and PADEP and other State agencies is the radon hazard. This hazard has the potential for large numbers of lung cancer among natural gas customers. This issue, which has been ignored in the DEC's revised Draft Supplemental Environmental Impact Statement (NYDEC, 2011), must be addressed in a revised Impact Statement and before DEC issues any hydraulic fracturing drilling permits in the Marcellus shale.

Unlike present sources for natural gas, located in Texas and Louisiana, the Marcellus Shale is considerably closer to New York and Pennsylvania consumers. In addition, the radioactive levels at the wellheads in New York and Pennsylvania are higher than the national average for natural gas wells throughout the United States.

In a paper we recently wrote (RWMA, 2012), RWMA calculated the wellhead concentrations of radon in natural gas from Marcellus Shale, the time to transit to consumers, particularly New York City residents, and the potential health effects of releasing radon, especially in the smaller living quarters found in urban areas.

It is well known that radon (radon-222) is present in natural gas (ATSDR, 1999) (NRC, 1988) and has been known since the early 1900's (Gesell, 1975). Published reports by R. H. Johnson of the U.S. Environmental Protection Agency (Johnson, 1973) and C. V. Gogolak of the U.S. Department of Energy (Gogolak, 1980) address this issue in depth, as we discuss below. Radon is present in natural gas from Marcellus Shale at much higher concentrations than natural gas from wells in Louisiana and Texas.

Since radon is a decay product of radium-226, to calculate radon levels it is necessary to know the concentrations of radium-226. Based on a USGS study (Leventhal, 1981) and gamma ray logs (also known as GAPI logs) that we have examined, as we mentioned earlier, the radium concentrations in the Marcellus Shale is 8 to 32 times background.

Using this range of radium concentrations and a simple Fortran program that simulates the production of radon in the well bore, and transit to the wellhead, we calculate a range of radon concentrations at the wellhead between 36.9 picoCuries per liter (pCi/L) to 2576 pCi/L. The maximum calculated concentration in Marcellus shale, 2576 pCi/L, is higher than the maximum concentrations found in the Panhandle region of Texas, 1450 pCi/L. (Johnson, 1973) (Gogolak, 1980).

These wellhead concentrations in Marcellus shale are up to 70 times the average in natural gas wells throughout the U.S. The average was calculated by R. H. Johnson of the U.S. Environmental Protection Agency in 1973 (pre-fracking) to be 37 pCi/L (Johnson, 1973), ranging between 5 pCi/L and 1450 pCi/L.

In addition, the distance to New York State apartments and homes from the Marcellus formation is 400 miles and sometimes less. This contrasts with the distance from the Gulf Coast and other formations, 1800 miles. At 10 mph movement in the pipeline (Gogolak, 1980), natural gas containing the radioactive gas, radon, which has a half-life of 3.8 days, will have three times the radon concentrations than natural gas originating at the Gulf Coast, everything else being equal, which it is not.
Being an inert gas, radon will not be destroyed when natural gas is burned in a kitchen stove.

We have examined published dilution factors and factored in numbers for average urban apartments where the dilution factor and the number of air exchanges per hour are less than those of non-urban dwellings. This analysis implies that the radon concentrations in New York City and urban apartments will be greater than the national average.

We assume 11.9 million residents in New York State are affected. This figure is calculated in the following manner: Based on U.S. Department of Energy figures our calculations assume 4.4 million gas stoves in New York State (EIA, 2009). This figure is multiplied by 2.69 persons per household to determine the number of residents affected: this number equals 11.9 million.

We calculate the number of excess lung cancer deaths for New York State. Our results: the potential number of fatal lung cancer deaths due to radon in natural gas from the Marcellus shale range from 1,182 to 30,448.

This is an additional number of lung cancer deaths due to radon from Marcellus Shale over deaths from natural radon already impacting New York State homes and their residents.

The Draft Supplemental Environmental Impact Statement produced by the New York State Department of Environmental Conservation (NYDEC, 2011) needs to be revised to take into account this public health and environmental hazard. In the entire 1400 page statement there is only one sentence containing the word "radon" and no consideration of this significant public health hazard.

Further, NYDEC should independently calculate and measure radon at the wellhead from the Marcellus Shale formation in presently operating wells before issuing horizontal hydraulic drilling permits in New York State. The presence of processing plants and storage must also be taken into account, and can serve to reduce the radon concentrations in natural gas delivered to consumers. The present rdsGEIS by DEC discusses none of these issues and should be rewritten.

CONCLUSIONS

Studies by the U.S. Geological Survey and gamma logs of drillers show radium concentrations up to 32 times background concentrations in the Marcellus shale formation. Brought to the surface in rock cuttings, drilling fluids, flowback water and brine, radium can enter the environment in several forms. During well development, rock cuttings and flowback water containing radium will be bought to the surface. Flowback water from horizontal hydraulic fracturing in the Marcellus formation has high concentrations of radium that must be disposed in a deep well or properly separated at a water treatment facility; the radioactive filters or sludges must go to a licensed landfill. During well production, radioactively-contaminated brine, up to 3000 times safe drinking water limits, must also be treated or disposed in a deep well. Radium-contaminated scale will form on the down well pipes as well as feeder lines and condensate tanks and these must be safely decontaminated; direct gamma from these pipes will increase the radiation dose to workers and the general public. Workers at pipeyards have developed cancer from direct gamma radiation and inhaling radioactive dust. The scale must be sent to a licensed disposal facility. Radioactive radon gas in natural gas from the Marcellus shale formation, much closer to the end use market than natural gas from Texas and Louisiana, will enter homes through kitchen stoves. Unless this natural gas is treated or held up, increased radon concentrations in homes and urban apartments will cause an increase in lung cancer rates.

REFERENCES

AAPG, Langhorne Smith, American Association of Petroleum Geologists 2010 National Conference, "Tectonic and Depositional Setting of Marcellus and Utica Black Shales," **2010**.

ATSDR, Agency for Toxic Substances and Disease Registry, U.S. Department of Health and Human Services, Public Health Service, *Toxicological Profile for Ionizing Radiation,* September **1999**, http://www.atsdr.cdc.gov/toxprofiles/tp149.pdf (accessed February 23, 2012).

Cornell, **2010**. "Gas Wells: Waste Management of Cuttings, Drilling Fluids, Hydrofrack

Water and Produced Water." Accessed on 24 March, 2010.
    <http://wri.eas.cornell.edu/gas_wells_waste.html>

Davies, Peter J., **2009**. "Radioactivity: A Description of its Nature, Dangers, Presence in the
    Marcellus Shale and Recommendations by the Town Of Dryden to The New York State
    Department of Environmental Conservation for Handling and Disposal of such
    Radioactive Materials." Cornell University.
    <http://www.tcgasmap.org/media/Radioactivity%20from%20Gas%20Drilling%20SGEIS
    %20Comments%20by%20Peter%20Davies.pdf>.2009

Energy Information Administration, United States Department of Energy, "Residential Energy
    Consumption Survey (RECS), RECS Survey Data, **2009**,"
    http://www.eia.gov/consumption/residential/data/2009/#undefined (accessed February 23,
    2012).

EPA, 40 CFR 192.12, **2011**

EPA, United Stated Environmental Protection Agency (US EPA), "A Preliminary Risk
    Assessment of Management and Disposal Options for Oil Field Wastes and Piping
    Contaminated with NORM in the State of Louisiana," RAE-9232/1-1, Rev.1, Prepared by
    Rogers and Associates and S. Cohen and Associates Inc. **1993**

Geology.com, **2008**, http://geology.com/articles/marcellus-shale.shtml

Gesell, T.F., University of Texas Health Science Center at Houston, "Occupational Radiation
    Exposure due to 222Rn in Natural Gas and Natural Gas Products," Health Physics,29, pp.
    681-687, **1975**.

Gogolak, C.V., Environmental Measurements Laboratory, United States Department of Energy,
    Review of ^{222}Rn in Natural Gas Produced from Unconventional Sources, (DOE/EML-
    385), **1980**.

Grove Software Incorporated, MicroShield, Version 8.02, **2008**

HPS, Health Physics Society, Relationship between mass and radioactivity,
    http://www.hps.org/publicinformation/ate/q6747.html, **2007**.

Hill, D.; Lombardi, T. E.; Martin, J., "Fractured Shale Gas Potential in New York," TICORA
    Geosciences, Inc., Arvada, Colorado, USA, **2004**.

Johnson, R.H. *et al.*, United States Environmental Protection Agency, Radiation Programs Office,
    Assessment of Potential Radiological Health Effects from Radon in Natural Gas,
    (EPA/520/73/004), **1973**.

Leventhal, J.; Crock, J.; Malcolm, M., United States Department of the Interior, Geological
    Survey, Geochemistry of trace elements and uranium in Devonian shales of the
    Appalachian Basin, (Open File Report 81-778), **1981**.

Myrick, T.; Berven, B.; Haywood, F., Oak Ridge National Laboratory, *State Background Radiation Levels: Results of Measurements Taken During 1975-1979*, (ORNL/TM-7343), **1981**.

NRC, United States National Research Council, *Health, Risks of Radon and Other Internally Deposited Alpha-Emitters: BEIR IV,* **1988**.

NYDEC, New York Department of Environmental Conservation, "Draft Supplemental Generic Environmental Impact Statement on the Oil, Gas and Solution Mining Regulatory Program." http://www.dec.ny.gov/energy/58440.html, **2009**

NYDEC, New York Department of Environmental Conservation, **2010**. Regulations: Chapter IV Quality Services. Accessed on 6 April, 2010. <http://www.dec.ny.gov/regs/2491.html>.

NYDEC, New York Department of Environmental Conservation, *Revised Draft Supplemental Generic Environmental Impact Statement on the Oil, Gas and Solution Mining Regulatory Program,* **2011**, http://www.dec.ny.gov/energy/75370.html (accessed February 23, 2012). p. 6-202

NYSERDA, New York State Energy Research Development Authority, "Technical Consulting Reports Prepared In Support Of The Draft Supplemental Generic Environmental Impact Statement For Natural Gas Production In New York State," by Alpha Geoscience, September **2009**.

NYTimes, New York Times, "Regulation Lax as Gas Wells' Tainted Water Hits Rivers," February **2011**.

RWMA, Resnikoff, M, "Radon in Natural Gas from Marcellus Shale," June **2012**.

USGS, Rowan, E.L., *et al,* "Radium Content of Oil and Gas Field Produced Waters in the Northern Appalachian Basin (USA): Summary and Discussion of Data," USGS Scientific Investigations Report 2011-5135, **2011**.

TREATMENT OF WATER FROM FRACTURING OPERATION FOR UNCONVENTIONAL GAS PRODUCTION

Paul T. Sun, Charles L. Meyer, Cor Kuijvenhoven, Sudini Padmasiri, and Vladimir Fedotov

Shell Oil Company, Houston, TX

*Email: Charles.meyer@shell.com

ABSTRACT

A general introduction of tight gas field fracturing technologies and their interactions with the water management is presented to set the framework for water treatment. Typical water quality of these produced waters is introduced first. The treatment requirement of the flowback (FB) and produced water (PW) are determined by its final disposition: reuse, deep well disposal, or surface water discharge. Each alternative demands different treatment.

The treatment can include any of three stages of treatment which we have termed primary, secondary, and tertiary. Primary treatment consists of one or more of the following: oxidation, coagulation and separation of different phases and sludge dewatering. The secondary treatment process includes sulfate and/or carbonate precipitation of divalent cations such as Ca^{++}, Ba^{++}, Sr^{++}, and Mg^{++}. Also important in this precipitation is the fate of the natural occurring radioactive material, mostly radium element. The tertiary process consists of desalination TDS removal via one or more desalination technologies, such as reverse osmosis or thermal evaporation. And finally, the disposal of the final brine either through deep well disposal or a crystallizer is also discussed.

The purpose of this paper is to describe the applicable water treatment technologies along with a listing of the pros and cons of each in specific applications.

KEYWORDS: Tight gas fracturing, water treatment, flowback and produced water quality, the treatment requirement, oxidation, coagulation, separation, sulfate and carbonate precipitation of the divalent cations, reverse osmosis, vapor recompression evaporator, thermal crystallizer and deep well disposal

INTRODUCTION

Unconventional gas, also called tight or shale gas, is the newly developed natural gas resource extracted from the low permeability soft shale rock layers. These reservoirs, if fully developed, can significantly increase the natural gas production in the US. Based on one estimate, it can extend the supply of the nation's gas by almost 100 years. (1)

Cost effective horizontal drilling and hydraulic fracturing are the two innovations that open up these tight gas formations for this game-changing production. Figure 1 illustrates the development of the combination of these technologies.

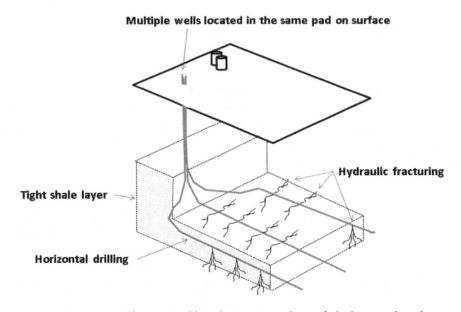

Figure 1. Simple presentation of shale gas development

Horizontal drilling develops the conduits to reach the majority of the gas-containing shale formation and the hydraulic fracturing artificially improves the surrounding permeability for the release of the gas, or sometimes oil, to flow to the well head.

The hydraulic fracturing process uses water under high pressure to open up the shale formation into fractures and pumps proppants (sand-size particles) in place to maintain the openings when water is withdrawn. Currently, two different water based fluids or a combination of those two (hybrid) are being used: slick water or cross-linked fluid.

The slick water frac uses large volumes of water, usually 3 to 4 million gallons of water per well, with added friction reducer to perform the fracturing and carry the proppants to the created openings. The process depends on the high velocity of the liquid and simpler chemistry. The friction reduction is accomplished by the addition of high molecular weight polyelectrolyte, usually several hundred ppm of either cationic or anionic polyacrylamide. Other possible components of the frac water are biocides, surfactants, clay stabilizers, and scale inhibitors. The addition of these components is not universal and depends on site-specific needs. The polyelectrolyte friction reducers tend to be adsorbed by the formation and the withdrawal of the frac water is easy, mostly without the addition of breakers.

The cross linked frac, on the other hand, uses lower molecular weight linear gel to reduce the friction of the water when fracturing the formation and later uses cross linking to thicken the gel for carrying the proppants to the proper locations. After the proppant placement, the thick gel is "broken" by the added slow-acting oxidants to break the cross-linkages and make it fluid again for easy withdrawal. Typically, guar gum is used as the linear gel. They are also cross-linked with borate at high pH and sometimes other similar chemicals are used as substitute. The water volume requirement of the cross linked frac is lower, about 1 million gallons total usage of water per well. Other components, such as biocides, surfactants, etc., are also added based on site specific needs. After fracturing, flow back water returns to the surface with the majority of the broken down gel material.

The industry is still developing the understanding of the frac water chemistry and its interactions with the geophysics of the formation. New formulations of different frac chemistries are constantly under development. The situation is very dynamic and the quality requirement for the makeup water is also continuously changing. (2) This complicates the water management for tight oil/gas development.

FLOWBACK WATER AND PRODUCED WATER

Once the stimulation (or fracturing) of the well is complete, the injected water will be produced back to the surface and eventually hydrocarbons will follow. Figure 2 shows a conceptual view of the water flow back development from a newly fractured well. In the beginning, a significant amount of water is released back to the well head without the presence of either the gas or liquid hydrocarbons. After several days to weeks of this flow back, oil and/or gas emerges. Logistically, the well completion team will clean up the site and hand over the well to Production Operations. By convention, the wastewater produced prior to this demarcation is called flowback water (FB) and afterwards, it is called produced water (PW). Although flowback water usually is a larger volume (10 to 30% of the injected water volume) and contains less salt content when compared with the produced water, it is actually the continuation of a natural water release process. The water production rate will gradually decrease to a lower level but it can continue for years with daily flow of 30 to 500 gallons per well depending on the local geological formation. The overall long term water recovery can be 30 to > 100% of the injection water.

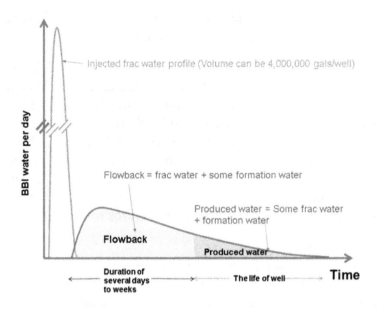

Figure 2. The injection water and flowback/produced water from a single well development

The characteristics of the FB/PW also change with time. Figure 3 demonstrates these variations. In general, the concentrations of the dissolved minerals, including Na^+, Ca^{++}, Mg^{++}, Ba^{++}, Sr^{++}, Mn^{++}, Fe^{++}, Si, B, Cl^-, and $SO_4^=$, solublized from the formation, increase with time and finally reach levels characterized by the geochemical properties of the formation. Typical total dissolved solids (TDS) concentrations range from 10,000 up to 300,000 mg/L. The concentration of components originated from the frac chemicals, such as broken guar gum and surfactants tend to decrease as the flowback water turns into produced water. Other important components such as total suspended solids or oil in water (when liquid hydrocarbons are part of the products) vary randomly with time.

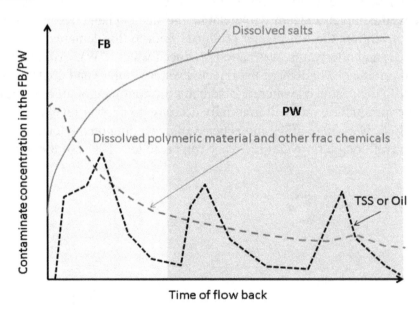

Figure 3. The general patterns of contaminant concentrations in the flowback/produced water

64

Table 1 lists typical characteristics of flowback and produced water. The ranges of dissolved minerals are highly variable and depend on the regional shale geochemistry. In addition, dissolved radionuclides can be present in these waters. The natural occurring radioactive material (NORM) concentration in some regions can be very high. (3) The organic material in the FB/PW consists of the residue of the polymeric material present in the frac chemical makeup, organic acids and dissolved aromatic hydrocarbons originating from the formation.

The flowback composition from a cross-link fracturing operation can be very different from a slick water frack. A large percentage of the broken guar gum is usually released back to the FB/PW and it can be results in a TOC level in the thousands of mg/L. This can contribute to high biological activities and thus the high bacterial counts in some samples.

Table 1. Typical Ranges of Components Present in Flow back/produced water

Parameters, mg/l	flowback water (first 2 weeks)	Produced water (> 2 week)
TDS	5,000-250,000	10,000-336,000
Sodium	2,000-100,000	4,000-135,000
Potassium	0-750	0-1,000
Calcium	0-20,000	0-40,000
Magnesium	0-2,000	0-4,000
Barium	0-10,000	0-20,000
Strontium	0-5,000	0-10,000
Iron	0-100	0-200
Chloride	3,000-150,000	6,000-200,000
Sulfate	0-1,000	0-5,000
Carbonate	0-1,000	0-1,000
Bicarbonate	100-6,000	100-6,000
Acetate	0-500	0-2,500
Propionate	0-100	0-400

Butyrate	0-25	0-75
BTEX	0-100	0-100
Radionuclide's	No reliable data available	
Specific gravity	1.000-1.250	1.000-1.250
Dissolved oxygen	0	0
Ammonia	10 - 200	10 - 200
Dissolved H_2S	0-1,000	0-1,000
pH	4-10	4-10
Temperature °C	20-150	20-150
Total suspended solids, mg/l	1-500	1-500
Oil-in-water, mg/l	5-1,000	5-1,000
Bacteria (total) per ml	$0-10^{10}$	$0-10^{10}$

LIFE CYCLE WATER BALANCES IN A GAS FIELD

In the development of the whole gas field, the overall fracking water demand and wastewater production generate an interesting life-cycle balance. An example is given in Figure 4. In this figure, no water recycle was assumed and only fresh source water is used for the fracturing. As one can see, in the beginning of a field development, the frac water requirement outpaces the flowback/produced water (FB/PW). The opportunity for reuse is high. Furthermore, the likelihood of meeting the frac water quality requirement is high using the fresh source water for diluting the FB/PW due to the typical low flowback/frac water ratio. The disposal or discharge of wastewater can be significantly reduced or eliminated by recycling. But as the field matures and well fracturing is slowing down, the opportunity for reuse of the FB/PW diminishes and finally, the disposal of the lower quality produced water becomes necessary. This back-end loaded disposal/discharge need for the gas development challenges the water management team for cost-effective life cycle planning.

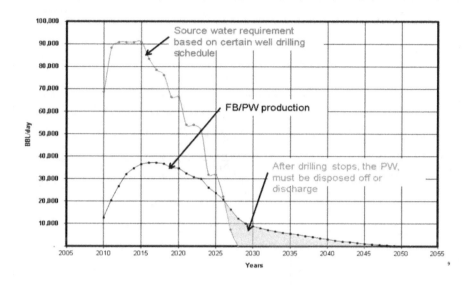

Figure 4.   Total water balance within a gas field

QUALITY REQUIREMENT FOR RECYCLE AND DISPOSAL

Like conventional produced water from oil and gas fields, there are only three routes that FB/PW can be disposed or reused: reuse as part of the fracturing water, discharge to receiving water, and disposal to a deep well.   Each has its own water quality requirement.   They will be discussed in the following sections:

Water quality requirements for reuse as part of the fracturing water- The required water quality differs significantly for the three different frac fluid categories as mentioned previously: slick water, crosslinked, and hybrid.   Hybrid fluids should follow the water quality guidelines for crosslinked fluids.   Furthermore, the on-going development of new fracking chemistry and technology will change the water quality requirement in the future.   Thus, the following water quality guidelines listed in Table 2 are provided for reference purposes only.   The real water uses will have to be approved by the technology vendors on site.

It is apparent from this table that cross-link frac technology demands significantly more stringent water quality for the reuse of the FB/PW than the slick water frac.   Unfortunately, different fracturing technology has its own application niche based on the geophysics of the field under development and the more stringent water quality will, therefore, be required for certain fields. Specific water treatment technologies, along with added freshwater for dilution, will be used to bridge the quality gap.

Table 2. Typical Water Quality Requirements for Frac Fluid Preparation

(In mg/L concentration)

Parameter	Cross –linked water	Slick water*
pH	6 to 8	none
Fe	<20	none
Total Hardness	<500**	none
Bicarbonate	<1000	none
Boron	<15	none
Si	<20	none
Sulfate	<50	none
TDS	***<40,000[1] or 70,000[2]	Up to 280,000
Bacteria count****	<100	none

*The situation is very fluid; the TDS requirement is dependent on the new polyelectrolyte development.
**This figure can be increased by adding anti-scalant
*** Depending on Zirconium[1] or Borate[2] cross-linking
****Total count per ml

Water quality requirements for direct discharge – The FB/PW contains high salinity and most of the shale fields in the U.S. are long distances away from coastal areas. Considering the current regulatory environment, it is believed that desalination of this wastewater is probably a minimal requirement in order to discharge to a receiving fresh water stream. As an example, a total dissolved solids (TDS) level of less than 500 mg/L is the current discharge limit required by the State of Pennsylvania. Limitations on other parameters would also have to be met on top of this stringent requirement. (4)

In addition to the limits on water quality, the disposal of brine concentrate or crystallized salt cake from the desalination operations can also be a major challenge, i.e., in some produced waters, their salt content can be 20 to 30% by weight of the wastewater flow. This massive solid disposal problem may change our thinking in the design the liquid treatment train. Due to this overall costly requirement, the surface discharge option usually is the last resort for final disposal of this wastewater.

Water quality requirements for deep well disposal - Subsurface disposal through deep well injection has been used by the oil industry as a safe, environmentally friendly and cost-effective option to dispose of wastes from E&P operations for several years now. In order to ensure that adequate injectivity conditions are maintained throughout disposal operations of the desired volumes, a number of considerations will have to be taken into account with respect to the quality of the waste fluids. These considerations will have to be made in the context of specific characteristics of the subsurface environment at hand such as:

1) Chemistry of the formation brine and rock mineralogy;

2) Reservoir temperature and pressure;

3) Rock permeability and pore structure.

The treatment requirement to prepare the FB/PW for subsurface injection would firstly be the removal of solids and oil to prevent plugging of the formation and secondarily modify the chemistry so that underground scaling would not happen, which would also plug the deep disposal wells. The specific limits of various ions will have to be dealt with individually.

GENERAL TREATMENT SCHEMES FOR THE FB/PW

The treatment requirement for the reuse of the flowback/produced water (FB/PW) is based on its characteristics and the required water quality demanded by the frac fluid with the consideration of dilution by the fresh source water. When the FB/PW production rate exceeds the demand for reuse, its treatment can be determined by the water disposal criteria, either surface discharge or subsurface disposal. In addition, the disposal of treatment residuals, such as sludge or salt cakes would also have to be dealt with.

Figure 5 outlines the general process in selecting the treatment schemes for FB/PW based on the final destination of the water.

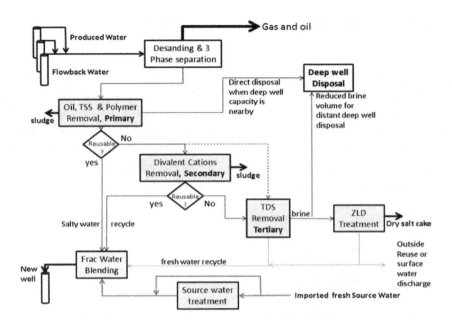

Figure 5. Water treatment in the overall water management scheme

Typically FB/PW contains some level of suspended solids, oil, residue gel material and precipitable ions, such as ferrous iron. These components can interfere with most of downstream unit operations and must be removed for all reuse or disposal purposes. The treatment train to accomplish this is termed primary treatment.

After this primary treatment, the effluent can be routed for deep well disposal or reuse if the frac water quality requirement can be met. In some cases, divalent cations (Ca^{++}, Ba^{++}, Sr^{++}, and Mg^{++}) removal may be required. Due to the high salt content and high divalent ion concentration, only chemical precipitation processes are useful in this application. They are called secondary treatment.

In some cases, TDS removal is required in order for frac reuse. Various desalination processes are applicable here. Some of the salt removal processes, such as membrane treatment, may require pretreatment to remove divalent ions while others may not. The overall desalination processes are called tertiary treatment.

The final disposal of FB/PW or the resulting brine from the tertiary treatment can either be by disposal into subsurface deep wells or by recovery of the water and salts separately by crystallization. These processes, without surface discharge of liquid wastewaters, are called zero-liquid-discharge or ZLD.

These treatment classification schemes are discussed in more details below.

PRIMARY TREATMENT FOR SOLIDS, OIL, RESIDUE POLYMER AND IRON REMOVAL

Primary treatment includes the removal of free oil, dispersed oil, suspended solids, iron, and the un-broken polymeric gel material. In addition, primary treatment also includes bacteria control. This primary treatment process is required for most FB/PW.

The main drivers for the treatment process selection are the presence of oil, suspended solids and residual polymeric material. A general process block-flow diagram for a primary treatment train is shown in Figure 6. It should be noted that the selection of the primary treatment process configuration will always be a project-specific decision as the combination of the feed water quality and required water quality in a specific project may not necessarily require all the shown steps.

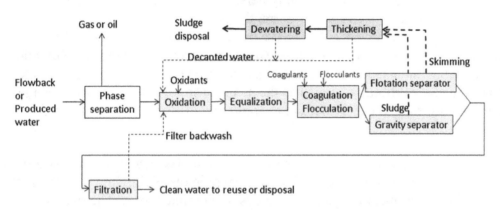

Figure 6. FB/PW primary treatment process block flow diagram

Oxidation - Chemical oxidation is usually required to oxidize the polymeric residual to smaller size molecules so that they lose the anti-coagulation properties and to oxidize ferrous to ferric iron for precipitation. Oxidants, such as sodium hypochlorite, ozone, chlorine dioxide, and even free radicals generated from advanced oxidation processes on-site have been used in the treatment of the FB/PW.

The oxidation reaction can be expressed as:

$$\text{Organics Others} + \begin{cases} \text{HClO (Chlorine)} \\ \text{ClO}_2 \text{ (Chlorine dioxide)} \\ \text{O}_3 \text{ (Ozone)} \\ \bullet\text{OH (Free radical)} \end{cases} \longrightarrow \text{Broken down organics} + \begin{array}{l} \text{Other end-products} \\ \text{(Chloramines, ferric,} \\ \text{MnO}_2, \text{S}^0) \end{array}$$

These reactions are kinetics limited and the rate of reaction depends on the oxidant used and the reductant species present in the water. After mixing the oxidants with the water, a certain minimum detention time is required for the reactions to complete. The detention time can be

71

provided by the flow time in transfer piping or in the equalization tank or even in the separators. Please note that some of the organic coagulants or flocculants will be attacked by these oxidants.

These oxidation processes perform the added benefit of disinfection for bacterial reduction and sulfide oxidation. The oxidants added also turn ferrous ions to ferric which precipitate and form a natural coagulant for oil and solids destabilization. This is probably the basis for some vendors in claiming the success of using oxidation treatment only for FB/PW cleanup without coagulant addition.

Coagulation - The typical organic polymeric treatment chemicals successfully used in conventional produced water treatment tend not to function well for this type of produced water. From the literature and our limited experience, it is concluded that only inorganic coagulants, such as alum or ferric chloride or their derivatives, are successfully used in destabilizing the colloidal particles or oil droplets. When ferric ions are added to the water, similar chemistry is also followed to generate inorganic iron-based polymers. These coagulants with their extended derivatives are the most dependable coagulants currently used in the field.
Some specialized organic polymeric coagulants have been used successfully also. Due to the complex aquatic chemistry of these types of wastewaters, the general mechanisms of colloid formation and its destabilization, such as colloid surface charge neutralization, inter-particle bridging, or enmeshment are difficult to assign as the mechanism of destabilization. Field jar testing through trial-and-error and experience are the only reliable ways to select the right treating chemicals.

Electrocoagulation (EC), sometimes called electro-precipitation, has been used successfully in treating wastewaters. This process uses direct electric-current in dissolving iron or aluminum electrodes in generating the primary coagulants– ferrous or aluminum ions. The freshly dissolved metal ions are claimed to be better destabilizing chemicals then the chemical forms pumped into the wastewater such as alum or ferric chloride. This EC process may be more suitable in treating the unconventional FB/PW due to the high salt concentration because of the high electrical conductivity. In the electrolysis process, hydrogen and sometimes chlorine or oxygen gases are also generated. These gases will assist the downstream flotation separation processes. (7), (8)

In the application of EC to the FB/PW treatment, most vendors have found that using electrocoagulation alone would not destabilize the colloids in this type of wastewater quickly; a pH elevation to 9 or 9.5 by adding caustic was deemed necessary in clarifying the suspended solids and oil.

Solids-liquid separation treatment - The destabilized flocs should be separated in a flotation separator if oil is present. The removed oil and solids in the form of sludge or float are to be thickened to reduce the free water content associated with them and then dewatered to dryer paste for landfill disposal. The oil associated with these waste streams is too little and difficult to recover. No slop oil recoveries are attempted for this unconventional produced water.

72

The induced gas flotation (IGF) equipment, normally used in conventional produced water treatment, does not function well in this unconventional application. The bubble size in these units are very large (>1000 μm) which are not ideal for solid/liquid separation

The micro-bubble flotation units (DGF or the dispersed gas flotation) are better separators due to the small gas bubble generated (usually < 100 μm), their less turbulent internal fluid movement and the available mechanical scraper mechanism for skim or float removal. (9)

For gravity settling, in cases where no oil is present in the feed stream, general gravity clarifiers, contact settlers with sludge recycling, and lamella clarifiers have all been used successfully. The selection of which type to use is dependent on solids loading, and its cost-effectiveness. Units with sludge recycle tend to produce better sludge for dewatering.

The treated effluent from either the flotation or settling unit is normally not good enough for disposal or reuse and furthermore, it may be contain too much suspended solids to be removed by a cartridge filter. A media filtration device is used to further polish the solids and/or oil. Media filters, such as sand filters or multi-media filters have been used successfully. The typical application factor is in 3 to 8 gpm/ft^2 of filter cross-sectional area. The nut-shell filters, normally used in conventional produced water treatment, have been proven to be problematic in this application. Apparently, the polymeric material in the wastewater would coat the nut shell media and render it difficult to backwash.

Sludge dewatering - The sludge or float withdrawn from the separators may be too wet to be either sent for direct disposal or dewatering. A sludge thickening tank, usually gravity separation, is used to remove free water from the solid mixture. The thickened sludge can then be dewatered to dryer cake for landfill disposal. Typical dewatering units used in this service are: dewatering roll off boxes, centrifuges or plate-and-frame sludge presses. Additional polymer addition may be required to condition the wet sludge for dewatering (10).

SECONDARY TREATMENT FOR DIVALENT IONS REMOVAL

Because of the typically high concentrations of divalent ions in FB/PW, chemical precipitation is the only feasible process in this removal category. Other removal processes, such as ion exchange or nanofiltration are not practical for this application. Usually carbonate (softening) or sulfate precipitation processes are used.

Typically the precipitation equipment described below can reduce the total hardness concentration to less than 100 mg/L as $CaCO_3$. This level of treatment more than meets the criteria for reuse of the FB/PW for frac water uses. However, certain tertiary and ZLD water treatment processes may require even lower levels of calcium or barium in feed water.

Carbonate precipitation - The removal of Ba^{++}, Ca^{++} and Sr^{++} from FB/PW is usually accomplished by chemical precipitation. It can be done by carbonate precipitation and Mg^{++} can also precipitated out of solution by forming hydroxide at high pH. This is easily carried out in a typical softening process where sodium carbonate and sodium hydroxide are added into a mixer/flocculator/settler for precipitation and clarification (11).

Under higher pH conditions (> 8.5) the conversion of bicarbonate to carbonate will induce the precipitation of calcium, barium and strontium salts. In addition, Mg^{++} ions will also precipitate under higher pH conditions (>10). Silicate can also be removed by adsorption onto magnesium hydroxide flocs in this series of reactions. Beyond the equilibrium predictions, there are practical limitations in applying these calculations in the design of the multiple cationic precipitation treatment systems. The final outcomes are controlled by the kinetics of these competing reactions and the presence of surfactants or anti-scalants in the FB/PW can slow down the kinetics of precipitate formation significantly.

The chemical precipitation reactions are carried out in a reactor with three zones – a rapid mixing chamber where chemicals are mixed with the feed water, a precipitation reactor where gentle mixing is provided to promote the particle growth, and a settling tank where the precipitated solids are separated from the treated water. Figure 1 shows a typical unit process train for this treatment. Some manufacturer's unit provides sludge recycle to send precipitated solids to the rapid mix tank to supply seeds for particle sizes to grow faster to improve reaction rate and good sludge settling and dewatering. With significant sludge solids production, the sludge processing costs can be high. Good dewatering is a key to lower the operational cost. The final clarified effluent will be pH-adjusted by adding acid.

Sulfate precipitation - The sulfate precipitation chemistry can be applied for the removal of Ca^{++}, Ba^{++}, and Sr^{++}. $BaSO_4$ precipitation is especially favored thermodynamically. However, both Sr^{++} and Ca^{++} ions will also be removed in sulfate precipitation, although not be to the same extent as Ba^{++}. For example, for calcium removal, the lowest concentration that can be achieved by sulfate precipitation is around 1,500 mg/L as Ca.

The reactions can be carried out in the same reactors as carbonate precipitation under most pH conditions. The chemical added is usually sodium sulfate. The growth of larger crystals is also promoted by recycling the settled solids back to the reaction zone and keeping the precipitated solids in the system for a long period of time (12). Sometimes using combined sulfate and carbonate precipitation to remove these divalent ions is the most appropriate method considering the reaction equilibrium and sludge disposal issues. These reactions can be promoted in similar reactors as shown in Figure 7.

Radionuclide Issues in Chemical Precipitation Treatment – Some FB/PW may contain dissolved radionuclide ions, predominantly divalent ions of Ra-226 and Ra-228. During the precipitation treatment, radium ions co-precipitate with barium sulfate:

$$Ba^{++} + Ra^{++} + SO_4^{=} \longrightarrow BaRaSO_4 \ (s)$$

In this way, the radium ions can be concentrated in the sludge, which may complicate the sludge handling and disposal. (13)

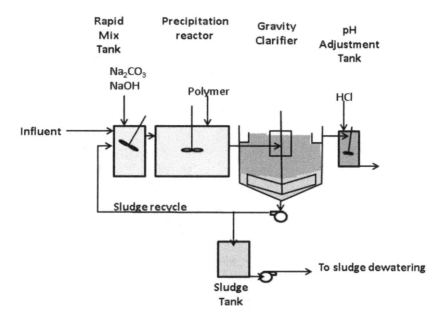

Figure 1. Carbonate Precipitation Process Flow Diagram

TERTIARY TREATMENT FOR DESALINATION

Tertiary treatment describes the removal of total dissolved salt from the FB/PW for reuse and disposal. This desalination treatment can involve reverse osmosis membranes, thermal distillation and other processes. The purpose is to produce freshwater with a residual of concentrated brine water for disposal or further drying. The philosophy here is to recover as much freshwater without forming fouling crystals either on the membrane of the Reverse Osmosis (RO) systems or the heat exchanger surface of the evaporators.

Selection of desalination technology primarily depends on the salt concentration of the influent water.   Figure shows the application range of these technologies (14). The concentration of the TDS is based on the assumption that NaCl salt is the only constituent in the feed water. Other divalent salts may influence the selection. However, they are a secondary factor which can be dealt with in the pretreatment steps.

For the desalination processes, reverse osmosis and thermal evaporation are the most applied in the field. They are discussed as below.

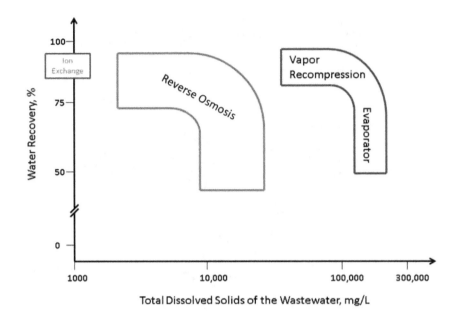

Figure 8. Application Windows of Desalination Technologies

Reverse Osmosis - The RO process is based on applying high inlet pressure to overcome the osmotic pressure of the brine water so that water molecules can be pushed through the semi-permeable membrane, whereas salt molecules are rejected. However, the system is limited by the pressure that can be economically applied to the membrane bundle. In general, it is limited to a feed water salt concentration not more than 40,000 mg/L. As the salts get concentrated in the process, the water recovery is reduced to accommodate the upper pressure limit. As other multi-valent salts may precipitate at lower water recovery condition, they would need to be removed in a secondary treatment unit preceding the TDS removal step. *A general process flow diagram is given in Figure 9.* Even with all these pretreatment steps, the system has to be further protected by lowering the pH and adding anti-scalant to protect the expensive membrane modules.

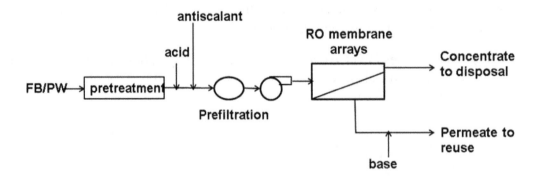

Figure 9. Reverse Osmosis Treatment System

76

Normally, a freshwater of less than 500 mg/L of total dissolved solids can be produced by the RO process treating FB/PW. However, there are some important components that may not be easily rejected by a normal RO process. One of them is boron and at neutral pH, boron rejection is less than 50%. So if B removal is important, an additional treatment process may be needed.

Mechanical Vapor Recompression Evaporator (Brine Concentrator) - Vapor Recompression Evaporation can be applied to treat more concentrated feed water. Typically for the small flow applications in the shale gas field application, the vapor recompression evaporator is more economical and has been applied successfully in the Barnett Play (15). The limit of applying this technology is the precipitation of salt, in this case NaCl. When the salt concentration in the brine reaches close to 250,000 to 300,000 mg/L, there is a danger of scaling up the heat exchanger surface of the evaporator by the NaCl crystals. When the FB/PW TDS concentration is in this range, an evaporator may not be workable and only a thermal crystallizer can be applied to produce both fresh water and salt cake for direct disposal.

A mechanical vapor recompression (MVR) evaporator uses vapor recompression by a gas compressor as the main driver for water evaporation - see Figure 10. It takes advantage of the principle of reducing the boiling point temperature by reducing the pressure at the vapor space. (16) The compressed vapor is then condensed in the shell side of the falling film heat exchanger at a higher condensation temperature where the latent heat is transferred to the brine circulating through the internal tubes. Its application range is from 20,000 to 250,000 mg/L of NaCl concentration.

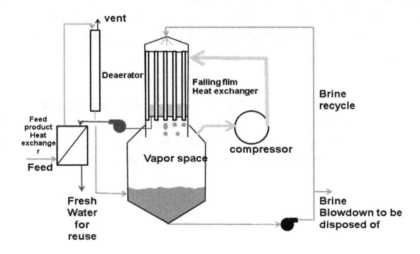

Figure 10. Schematic Diagram of a Mechanical Vapor Recompression evaporator

Although the pretreatment requirements for a MVR evaporator are not as stringent as those for the RO system, we still recommend that some divalent cations removal (secondary treatment) be applied as a pretreatment step to improve the freshwater recovery and scaling prevention for the application in FB/PW treatment.

ZERO LIQUID DISCHARGE TREATMENT

The objective of zero liquid discharge is to eliminate any liquid waste at the end of the water treatment process. Evaporators or concentrators can be utilized to concentrate the waste stream, but the conversion of the concentrated brine into solids/salts form must be accomplished by using a crystallizer. Disposal of solid waste from the crystallizer must avoid contamination of surface or groundwater.

In order to totally eliminate surface discharges, the most cost-effective method has been deep well disposal if the wells have capacity and are nearby. Without the available deep disposal wells, one can only evaporate the water away from the brine and produce dry salt cakes for secure landfill disposal. Other natural disposal processes, including solar ponds or Freeze/Thaw/Evaporate (FTE), are geographically restricted to arid and/or cold environments so they will not be discussed further.

Thermal Atmospheric Crystallizer - Since the thermo-evaporators cannot crystallize large quantities of dissolved salts, a thermal atmospheric crystallizer is used to further concentrate the inlet brine from 200,000 to 300,000 mg/L of salt as NaCl (or 20 to 30 % by weight) to recover the last bit of water and produce salt cake (16). The forced circulation crystallizer is commonly used for this purpose.

As shown in Figure 11, the brine with precipitated solids slurry is forced to circulate through a brine heater where steam is used to input the heat for evaporation. Here, due to the higher pressure, the brine stream is heated up but no evaporation occurs. The heated brine is released into the crystallizer vapor body where water is vaporized and salts are precipitated. The water vapor is then condensed through a heat exchanger where cooling water is used to condense the vapor into clean distillate for reuse.

Part of the brine/slurry stream from the crystallizer vapor body is sent to a solids removal / solids dewatering unit where the precipitated salts are released from the system as salt cake for final disposal. Although the energy for water evaporation is high here (approximately 1 lb of steam for 1 lb of water evaporated), the system is easy to construct and operate.

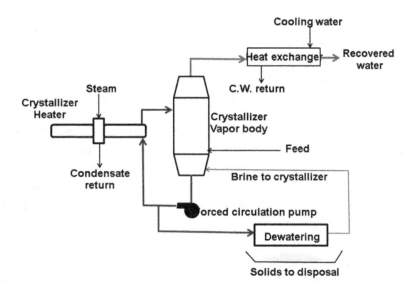

Figure 11. Forced-circulation evaporator-crystallizer

The working philosophy of an evaporator is opposite to that of the crystallizer. For easy operation of an evaporator, one needs to keep concentrations of easily precipitating salts as low as possible to protect the heat exchanger from scaling. These can be sulfate salts of Ba, Sr, or silicates. For easy operation of a crystallizer, on the other hand, one needs to have the easily precipitating salt concentrations as high as possible. These include NaCl which precipitates out at about 25 to 30 wt %. This will allow the crystallizer operate without having to deal with high boiling point elevation.

Alternatively, one can take the last volume of the concentrate brine and store it in an evaporation pond for natural drying if the local climate is conducive for such a system. Or one can follow the crystallizer with a spray dryer. But the energy use for the dryer can be as high as 230 KWH per bbl of water evaporated – a large amount of energy for recovery of a small amount of water. Other ways of getting rid of this difficult to deal with brine is to use it as drilling fluid or fix it by using fly ash to solidify this brine. All of these alternative ways for final brine disposal have limitations and challenges.

SUMMARY

1. Compared to conventional oil & gas fields, the unconventional produced water quality is influenced not only by the formation, but also by the fracturing fluid *introduced* to the formation. Water/rock interaction is playing a major role in determining the FB/PW water quality. As a result, produced fluids may contain a combination of hydrocarbons, dissolved inorganic salts, organic acids, residual

fracturing fluid additives and daughter products of their decomposition, mobilized formation solids, proppant, scales, corrosion products and biomass.

2. Typically the quality of raw produced/flowback water makes it unsuitable for fracturing, disposal or beneficial use without some degree of clean-up treatment. Treatment of water of highly variable quality with a robust process in a cost-effective manner presents specific challenges. The conventional oilfield water treatment technologies may not be adequate due to the presence of specific constituents of concern, which are critical for re-use for fracturing or disposal.

3. Salt tolerance of the frac fluid formulation is the main driver that determines the feasible extent of FB/PW re-use. Continuous technological advances in frac fluid formulations increase the tolerance to dissolved salts, leading to relaxation of the water quality requirements, which allows the use of produced brines for fracking without costly salt removal treatment

4. The treatment processes for unconventional oil/gas FB/PW can be divided into four functional categories, listed below in the order they are likely to be combined when so required:

   - Primary treatment - removal of suspended solids, condensate, polymers and iron as well as bacteria control to prevent fouling and souring effects at surface and subsurface and interference with fracturing fluid performance
   - Secondary treatment - removal of specific cations such as Ca^{2+}, Mg^{2+}, Ba^{2+}, Sr^{2+} to satisfy the ionic compatibility of water with fracturing fluid additives, scale control or as a pre-treatment step for tertiary and ZLD treatment
   - Tertiary treatment - to produce freshwater (TDS < 500 mg/L) for re-use for fracturing or other beneficial use and a residual of concentrated brine for disposal or further drying
   - Zero Liquid Discharge (ZLD) treatment – to separate highly concentrated brine into a dry salt cake for disposal and fresh water stream for re-use for fracturing or other beneficial use

REFERENCES

1. US Department of Energy, Modern shale gas development in the US: a primer, April, 2009.
2. Rickman, R., M. Mullen, E. Petre, B. Griser, and D. Kendert, "A practical use of shale petro-physics for stimulation design optimization: All shale plays are not clones of Barnett Shale", Halliburton, SPE 115258, 2008.
3. Unpublished data of Pennsylvania Department of Environmental Protection (2009-2010), cited in: "Radium Content of Oil- and Gas-Field Produced Waters in the Northern Appalachian Basin (US): Summary and Discussion of Data". Scientific Investigations Report 2011–5135. U.S. Department of the Interior, U.S. Geological Survey
4. Pennsylvania Bulletin, Volume 40, Number 34, August 21, 2010.
5. Baker, B.D., et al., "A Balanced Approach to Waste Disposal using UFI (Underground Fracture Injection), SPE 71433

6. Ely, J. W. Horn, A., Cathey, R. and Fraim, M. "Game changing technology of treating and recycling frac water", SPE-145454-PP, 2011

7. Holt, P. K and et.al., "The future for electrocoagulation as a localized water treatment technology", Chemosphere, 59, 355-367, 2005

8. James Edzwald and Johannes Haarhoff, "Seawater coagulation for reverse osmosis: chemistry, contaminants and coagulation", Water Research, Vol. 45, P. 5428, 2011.

9. Dufour, R. et. al., "Dissolved air flotation and the Poseipump, Paper 360°, September, 2006.

10. "Operation and maintenance of sludge dewatering systems", Water Pollution Control Federation Manual of Practice, 1987.

11. Bourcier, W. L., H. Brandt, and J. H. Tail, "Pretreatment of oil field and mine waste waters for Reverse Osmosis", Chapter 46 in Produced water 2 – environmental issues and mitigation technologies, Ed by Rark Reed and Stale Johnson, Plenum Press, 1995.

12. Keister, T. "Marcellus Shale gas well wastewater treatment needs", Paper presented at the Science of Marcellus Shale hydrofracture flowback and production wastewater treatment, recycle, and disposal, at Lycoming College, Williamsport, PA, Jan, 2010

13. Silva J., Matis H., and Tinto J., "NORM Removal from Frac Water in a Central Treatment Facility", Paper IWC 10-65, International Water Conference, 2010. Pittsburg, PA.

14. Gas Technology Institute, "Techno-economic assessment of water management solutions", 2011

15. Jevons, K. and Awe, M, "Economic benefits of membrane technology vs. evaporator", Desalination, Vol. 250, P. 962, 2010

16. Robertson, J., "Emerging technologies and challenges in water use and re-use in heavy oil industry", presented at the CHOA Technical Luncheon, October 2, 2007.

17. Jenkins, K, T. Higgins, J. Mavis, T. Sandy, L. Reid and K. Martins, "Conceptual design and evaluation of zero liquid discharge systems for management of industrial wastewater", presented at the 2011 Water Environment Federation Technical Exhibition Conference (WEFTEC), Los Angeles

CRITERIA FOR FLOWBACK WATER RECYCLE

John J. Schubert, P.E.

HDR Engineering

ABSTRACT

One of the major environmental issues associated with the hydrofracturing of shale gas formations has been the disposition of water flowing back from the well after hydrofracturing. The flowback water contains substantial amounts of dissolved salts, and can not be treated effectively using conventional treatment facilities. In the Marcellus Shale play, producers have turned to reusing the flowback water in future hydrofracturing events. Treatment levels vary from producer to producer. Some producers move flowback water directly to the next hydrofracturing event. Others provide varying levels of treatment. Many in the water treatment industry want to provide services to the industry, but don't understand the requirements of recycle water quality. This paper reviews the factors that limit the reuse of flowback water, and the progress made in recycling. Those factors include friction reduction, disinfection and scaling control.

KEYWORDS: Hydrofracturing, flowback reuse, criteria

Introduction

When my firm became interested in the water end of hydrofracturing, in 2008, the thought process was much different than today. The number of wells placed into the Marcellus Shale was very small. None of the producers were willing to share data, so not much was widely known about flowback and produced water characteristics, and much of the flowback water was going to sewage treatment plants for discharge without much removal of salts, metals, etc. As the thinking progressed, it was widely recognized that discharge in this manner could not continue. In Pennsylvania, in April 2011 the state took steps to discontinue the practice of discharging to municipal treatment plants. Evaporation/ crystallization facilities were proposed as early as 2009 and seriously considered, but the cost was felt to be prohibitive, and the management of gas producing companies required alternatives. By the end of 2011, over 90% of the flowback water from major producers was being recycled (1, 2). While this is commonly cited in industry publications and literature, there is little information provided on the constraints and limitations of this approach. This paper is provided to attempt to provide some background on the practice of flowback recycle. While this paper is based largely on experience in the Marcellus Shale, the rationale regarding frack water reuse is applicable in other shale gas plays.

For those that do not work in this environment, a few definitions are in order. Most know that in hydrofracturing, water is pumped into wells at a high pressure, and the shale formation that the water is pumped into fractures from the water pressure. When pumping is discontinued after fracture, some of the water flows back from the well as a stream that is mostly water with some gas. This continues for a few days, with the amount of gas increasing and the amount of water decreasing, until the point is reached where almost all of the stream is gas, and the well is put online. The water recovered from the well in those first few days when the stream from the well is mostly water is referred to as flowback. After the flowback period, water recovered from the well is referred to as produced water. The water in contact with the shale formation tends to dissolve soluble components that come into contact with it. The water

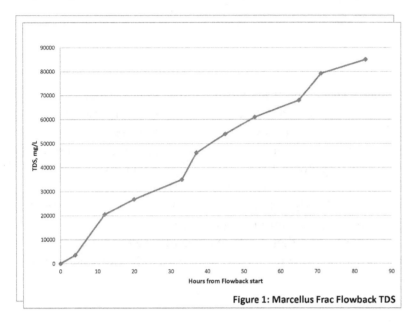

Figure 1: Marcellus Frac Flowback TDS

at the onset of flowback has been in minimal contact with the formation, and the dissolved salt pickup is relatively small. Figure 1 shows TDS concentrations in flowback water as a function of time from the start of flowback, from a typical Marcellus Shale well in southwestern PA. As shown, TDS levels increase from low levels initially to over 80,000 mg/L TDS in 80 hours. Final TDS concentrations at the end of the flowback period water generally exceed 100,000 mg/L. In comparison, produced water from the Marcellus Shale typically exceeds 200,000 mg/L TDS. While most of the TDS in the Marcellus Shale Play is sodium chloride, concentrations of parameters such as calcium, barium, strontium and magnesium vary substantially based on formation conditions, and can be somewhat related to location in the play.

In comparison to the TDS values shown in Figure 1, Figure 2 shows flowback flow characteristics from a representative Marcellus Shale well in southwestern PA. As shown, the flowback flowrate drops off significantly between day 4 and day 6. By day 12, the flowback water generated is less than 20,000 gallons/day, or a few gpm. It is not unusual for wells to be taken out of flowback after 4 to 5 days under current practices. Comparing the TDS concentrations and flowback flowrates shows that initially the well gives back

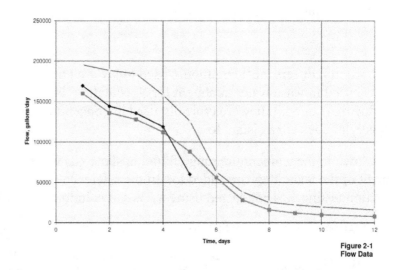

Figure 2-1
Flow Data

high flows of relatively dilute water, and as higher TDS levels occur, the flowrate drops. For this reason, the composite TDS level in flowback water is normally more on the order of 50,000 to 60,000 mg/L.

It has been widely observed that only a portion of the water applied to a hydrofracturing event returns as flowback water. While there has been considerable variation in the percentage of applied water that returns as flowback, the most often quoted values are in the range of 20 to 25% as average values. The author has observed values as low as 10% and as high as 40%, but generally in the vicinity of 20% to 25% is a reasonable expectation for horizontal wells. The amount of water applied depends on the number of stages in a hydrofracture and the amount of water required to achieve fracturing for each stage. Typical expectations for a horizontal Marcellus Shale Wells are 3,500,000 to 5,000,000 gallons applied. At a return rate of 20%, this produces between 700,000 and 1,000,000 gallons of flowback water.

In conversations with producers and service companies, the three concerns associated with reusing flowback water are (1) friction reduction, (2) biological activity and (3) scaling in the well.

Friction Reduction

Friction reduction refers to the addition of chemicals to reduce the friction loss of pumping at a high rate into the well. Literature and observation of past frack events indicates a typical pumping rate of 1600 to 4200 gpm (38 to 100 bbl/min) at high pressures into a 5 ½" OD carbon steel N80 production casing. This production casing with a ½" wall thickness is rated for a failure pressure of 13,220 psi (3), while a 0.361" wall thickness is rated for a failure pressure of 8830 psi. Marcellus Shale formation pressure is typically cited as about 4500 psi (4). Additional pressure is required to fracture the shale. Friction losses at 1600 gpm would exceed 4000 psi, assuming a 1 mile depth and a 1.5 mile horizontal run, and would get much larger at higher flowrates. Thus to maintain the ability to use commercially available and "bendable" pipe for shale gas wells, it is critical to minimize the friction loss in fluid movement. Friction reducers are added to frac water to accomplish this.

The friction reducers most commonly used today are polyacrylamide polymers. These polymers are also used in wastewater treatment for flocculation, but are effective at higher concentrations in reducing friction. They are typically used in concentrations of 500 to 1000 mg/L. Polyacrylamides are resistant to

breakdown in the temperature range found in shale gas wells (5), but are limited by the salt content of the water they are added to. In the reference cited previously, in late 2008, friction reduction testing was conducted using a 7% salt solution as the highest concentration. In that

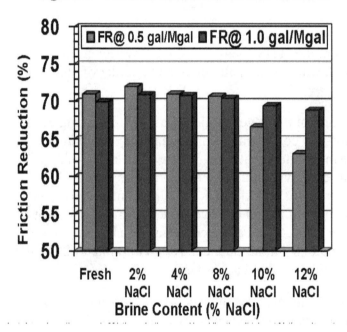

Figure 3 - Friction Reduction vs Salt Content

time frame, that was thought to be pushing the boundaries of salt content without affecting friction reduction. Oxidizing agents (typically persulfates) were and still are used to destroy polymer left adsorbed to the shale, which would otherwise tend to cover some of the surface of the shale and reduce gas production.

Proprietary products introduced since that time have shown higher tolerance for salt content, to above 12% TDS, as shown in Figure 3. Figure 3 is taken from a 2009 paper (6) and reflects the performance of a slightly anionic friction reducing polymer, which is tolerant to salt up to 12%. The friction reducer is advertised to degrade on its own, to not adsorb to the surface of the shale (due to its charging being about the same as the shale surface charge), and could be catalytically degraded if necessary. Others (7) have used proprietary polymeric friction reducers in frack water over 12% TDS with favorable results, which in part are possible because of pretreatment of flowback water for removal of divalent ions. Figure 3 does indicate that effectiveness starts to drop off at 8% NaCl, and at higher salt concentrations, a higher dosage may be needed to achieve comparable results. The industry rule of thumb target for friction reduction is 70% reduction. That is, the test result with friction reducer should be 70% less than pressure drop associated with no friction reducer. Conversations with producers in the past have indicated limitations on friction reducers resulting from calcium, magnesium, and iron, all of which interfered with friction reducer performance. Typically cited values above which interference with friction reducers was observed, albeit anecdotal, were typically a few ppm for iron, and 1000 mg/L each for Ca and Mg.

Friction reduction is typically evaluated using an arrangement called a friction loop. This device consists of a small tank, pump (frequently progressive cavity type) and small diameter tubing which mimics the velocity and Reynolds Number in the well production casing, and produces significant friction loss as a result. Initially water or a frack fluid is pumped around the loop without the addition of a friction reducer. Pressure is recorded at several points along the loop. Following that, the dosage of one or more friction reducers is varied, and pressures are again monitored. This approach is used to determine the dosage to be used in the field. Additionally, this type of apparatus allows for evaluation of possible impacts on friction reducers from biocides and other frack water additives.

The reuse of flowback water was limited early in the development of the Marcellus Shale by the tolerance of friction reducers for dissolved salts. With the development of new friction reducers that are less sensitive to salt content, this is less the case. The observed characteristic of Marcellus Shale wells as noted above is that about 20% to 25% of the water applied in a frackreturns as flowback. This indicates that even if the flowback and produced water from a given well together average close to 20% TDS, by the time they are diluted with fresh water 4/1, the resulting mixture will be only about 5% TDS, which is certainly tolerable from a friction reduction standpoint.

Disinfection
The second consideration involves biological activity potentially fouling the well. Principal bacteria of concern include sulfate reducing bacteria, anaerobic acid formers, and aerobic slime-forming bacteria (more of a concern above ground and in the well bore than in the shale formation)(8). While sulfates are generally low in the formation, there may be sulfates present in the fresh water source used for fracking. Organic substrate sources include friction reducers and other additives to the frack water from the frack water recycled for reuse. Primary concerns associated with biological activity include growth of acid forming bacteria in the well casing, which leads to corrosion of the casing, and formation of slimes and souring of the well from biological activity in the formation.(9)

Key criteria for evaluating disinfecting agents include effectiveness in killing bacteria, interaction with friction reducers, gels and other chemicals in the frac water, and environmental persistence. Another factor is the ability to measure a residual in the field to ensure effectiveness. While oxidizing biocides are generally effective, and easily measurable, interaction with other components in the water, including the aforementioned chemical additives as well as ammonia, may make them unattractive for downhole use. Disinfection of raw water supplies for fracking has been accomplished with chlorine dioxide and ozonation systems. Ozonation mobile systems have been used to degrade organic matter in flowback water as well as providing disinfection for flowback water.(10)

A number of non-oxidizing bactericides have been used in Shale Gas plays as a component of the hydrofracturing fluid. One commonly used biocide is glutaraldehyde. Glutaraldehyde has the chemical formula $CH_2(CH_2CHO)_2$. Other uses include disinfection of medical and dental equipment, and biological control in a variety of industrial water systems. Glutaraldehyde kills bacterial cells through cross-linking proteins. It does not interact with common friction reducers, antiscalants or other hydrofracturing water additives. Typical concentrations of glutaraldehyde in frack water are on the order of 500 mg/L. Colorimetric test kits are available from Hach and others for testing residual in the field. Other amine and carbamate non-oxidizing biocides are also used as downhole biocides.

Caution is required in selection of biocides due to possible interaction with the friction reducer. In the case of using anionic polyacrylamide as a friction reducer, use of cationic quartenary amines either separately or in combination with glutaraldehyde has been shown (11) to cause a reduction in friction reducer efficacy. Friction loop testing provides insight into the reactions between components such as biocides and friction reducers.

Contaminant Variability and Scaling

The third area of concern in the recycle of frac flowback water is that of scale-forming reactions in the formation. To appreciate the problem of scale control requires an understanding of the contaminants that accumulate in flowback water. Variations in flowback constituents occur from formation to formation, and even within formations. The Marcellus Shale provides several examples of variations within a single shale play. One variable in the Marcellus Shale is characterized in the oil and gas industry as liquids. Schlumberger defines natural gas liquids (NGLs) as components of natural gas that are liquid at surface in-field facilities or in gas processing plants. Natural gas liquids are classified according to their vapor pressure as low (condensates), intermediate (natural gasoline) and high vapor pressures. Natural gas liquids include butane, pentane, hexane and heptanes. Natural gas containing NGLs is referred to as wet gas, while natural gas with low or no NGLs is referred to as dry gas. To meet natural gas specifications for pipeline transport to use, NGLs must be largely removed, to achieve a consistent BTU content in the natural gas sold to users. In the Marcellus Shale, wet gas is observed from the northwestern portion of West Virginia through the western portion of Pennsylvania, as shown in the attached map. Butane and pentane boil at temperatures below the temperature of returning frac water, and all NGLs have limited solubility in water. Gravity separation of NGLs from the liquid phase and volatilization of low boiling NGLs limits the amount in flowback water for recycle.

Another variable, of potentially greater concern in recycling is the group of cations that form precipitate scale. Barium and calcium are of particular concern in flowback water. Barium in flowback water has been noted as varying from concentrations in the low hundreds of ppm in southwestern PA to over 5,000 mg/L in wells in northeastern PA. A well in Bradford County contained over 13000 mg/L Ba in late flowback water (12). Characteristics of mid-flowback or composite flowback water from wells in these areas is shown in the attached Table 1. Observation of the data indicates that barium can range from low values (just over 100 mg/L) to over 5000 mg/L. Calcium concentrations vary in the table from under 1000 to over 7500 mg/L. Similar variations have been observed in magnesium and strontium concentration.

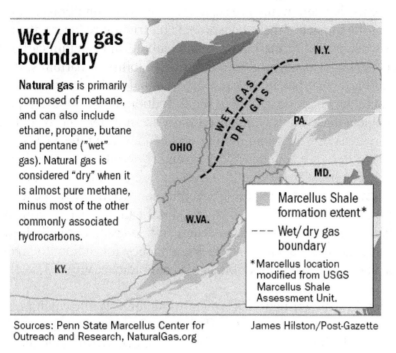

Wet/dry gas boundary

Natural gas is primarily composed of methane, and can also include ethane, propane, butane and pentane ("wet" gas). Natural gas is considered "dry" when it is almost pure methane, minus most of the other commonly associated hydrocarbons.

Marcellus Shale formation extent*

--- Wet/dry gas boundary

*Marcellus location modified from USGS Marcellus Shale Assessment Unit.

Sources: Penn State Marcellus Center for Outreach and Research, NaturalGas.org

James Hilston/Post-Gazette

Table 1

Scalant Composite Concentrations in Flowback Water - Marcellus Shale Play

parameter	Davis Well Marshall Co WV Day 5 flowback May 2009	Southwest Pa February 2009 Day 3+ flowback	Williamsport Pa Composite sample	Susquehanna Co Composite
TDS, mg/L	100,000	105,000	110,000	40,000
Ca, mg/L	7630	7500	6896	736
Mg, mg/L	829	640	725	127
Sr, mg/L		1700		228
Ba, mg/L	136	170	5145	596
Fe, mg/L	38	24	39	8

Comparing the constituents that commonly are scale forming with conditions in the well, and neglecting for a moment the impact of these constituents on friction reducers, the primary concerns for the well owner are the formation of barium sulfate and calcium carbonate in the well casing. Barium is the larger concern, due to the low solubility of barium sulfate ($K_{sp} = 1 \times 10^{-10}$). The solubility of barium sulfate in the presence of equal molar amounts is only about 2.4 mg/L $BaSO_4$. While there may be little sulfate In the formation, there is barium present. As barium dissolves, it contacts frack water which is partially fresh water, and would be expected to contain some sulfate. In areas in which coal mining was conducted in the past (i.e. most of the Marcellus Shale regions in Pennsylvania and West Virginia), sulfate concentrations in streams contaminated with AMD can be over 100 mg/L. Given an excess of barium in the formation and 100 mg/L sulfate in the frack water, the resulting precipitation reaction could cause almost 250 mg/L barium sulfate precipitation, which could have a significant impact in terms of coating the pores of the formation, and reducing gas flow. Barium sulfate has historically been observed as scale forming inside shale gas well casings. There is a tendency for radium to precipitate with barium in these scales, and they typically exhibit significant radioactivity (13).

The second consideration in scale formation is typically calcium. While calcium, and strontium can precipitate with sulfate (14), there is rarely sufficient sulfate present to react with all the barium in frackwater (after the initial day or so of flowback), even with moderately high fresh water sulfate concentrations. The exception to this would be the use of acid mine drainage for hydrofracturing, as that water can contain much higher sulfate concentrations, and would typically require some form of treatment to avoid barium and calcium sulfate scaling. Calcium is a key component of the calculation of most scaling indices, resulting from projected formation of calcium carbonate scale. However, the water used for fracking is usually near neutral pH. The Langelier Saturation Index (LSI) is routinely monitored to ensure scaling conditions are not provided in the feed water, and would be used if necessary to justify limiting the amount of recycle. Downhole, the pH of the frack water generally decreases. This may be from the soluble components of the formation, or from contact with CO2 in the natural gas stream (typically indicated as between 0.1% and 1%).

Service companies are typically careful in evaluating the chemistry of downhole interactions with frac water, and model the possible interactions to select anti-scalants and dispersants. As noted above, the limitations on calcium and magnesium for compatibility with the friction reducer are generally more severe than scaling considerations.

Environmental Considerations
Environmental considerations should also be noted in evaluation of frack water recycling. Double containment is typically required for storage of flowback water for reuse. The industry practice is normally storage in frac tanks on a contained pad. At least one producer has installed a double lined pond for storage of flowback water for reuse. In some merchant treatment facilities, permanent storage tanks in containment are provided for storage of both system feed and discharge. Permitting is required for temporary or permanent storage facilities

A spill of flowback water of as little as 5 gallons may be considered a reportable quantity. Permitting is required for storage facilities. Still, the flowback water has to be handled regardless of whether or not it is reused, and the risks of longer over the road travel are probably greater than the risks associated with storage for reuse.

Observations and Conclusions

Current practice in the field varies from producer to producer. There are facilities operating today in the Marcellus Shale play that provide pretreatment for reuse consisting of filtration and iron removal. Some remove barium as well through addition of sodium sulfate or sulfuric acid. Some provide oxidation for breakdown of organics, in fixed or mobile systems. Systems have been operated to also achieve some removal of calcium or magnesium, to meet producer requirements for those parameters. Unfortunately, there is not a fixed set of criteria that must be met. The requirements of producers vary based on the anticipated return flow of flowback water, and the requirements for compatibility with other fracking chemicals and potential scaling conditions.

While most producers are focused on diluting flowback water for reuse with fresh water, with flowback use in the vicinity of 25% of the total frack water supply, one producer recently reported hydrofracturing with 100% flowback water (15). The water was collected and treated to remove most of the barium and iron, and over half the magnesium. A smaller portion of calcium and strontium was removed. The frack was viewed as a success. The report did not describe the cost differential for friction reducers or other chemicals. In the author's past experience, water treated for barium, iron and magnesium removal was successfully reused diluted 1/1 with fresh water, with satisfactory friction reducer compatabilty and hydrofracturing results.

To summarize, the amount of flowback water that can be recycled depends on a number of factors, including compatibility with friction reducers and other chemical additives, concerns related to biological activity, and potential scaling reactions. Most water that is recycled requires at the minimum iron and barium removal and disinfection. More advanced treatment may be required by some producers to achieve compatibility with desired friction reducers or to meet downhole chemistry requirements. In that respect, the role of those with an interest in providing treatment of flowback water for recycle will generally be defined by the producer and other service providers in terms of acceptable water quality for recycle.

1. International Energy Agency, *Golden Rules for a Golden Age of Gas, World Energy Outlook Special Report on Unconventional Gas*, IEA, 2012, pp 32-35.

2. J.D. Krohn, "Water Recycling – It's Working in Pennsylvania", Energy in Depth, July 2011

3. V & M Tubes On-line catalog, 2011

4. Economic Comparison of Multi-Lateral Drilling over Horizontal Drilling for Marcellus Shale Field Development, Husain et al, Penn State University, January 2011.

5. SPE 119900 Critical Evaluations of Additives Used In Shale Slickwater Fracs, Kaufmann et al, 2008

6. SPE 125987Fracture-Stimulation in the Marcellus Shale – Lessons Learned in Fluid Selection and Execution, Houston et al, 2009

7. Gas Well treated with 100% reused frac fluid, Papso et al, Exploration and Production Magazine, 2010

8. SPE 123450 - Controlling Bacteria in Recycled Production Water for Completion and Workover Operations, Tischler et al, SPE Production and Operations Journal, May 2010

9. All Consulting "Water Treatment Technology Fact Sheet", 2010

10. See internet publications from Kerfoot Technologies, Fountain Quail Water Management LLC, Ecosphere and others.

11. SPE 119569 – Are You Buying Too Much Friction Reducer Because of Your Biocide, Rimassa et al, 2009, 2009 SPE Hydraulic Fracturing Technology Conference.

12. Silva, NORM Removal from Hydrofracturing Water, IWC 11-07, International Water Conference, 2011

13. Silva, NORM Removal from Frac Water in a Central Facility, IWC10-65, International Water Conference, 2010

14. Precipitation Reactions in Hydrofracturing Wastewater Treatment, Schubert, 2010 International Water Conference, IWC 10-64.

15. Gas Well Treated with 100% Reused Frac Fluid, Exploration and Production Magazine, August 2010, Papso et al.

BROMIDE IN THE ALLEGHENY RIVER AND THMS IN PITTSBURGH DRINKING WATER

Stanley States[1], John Kuchta[1], Georgina Cyprych[1], Jason Monnell[2], Mark Stoner[1],

Leonard Casson[2]* and Faith Wydra[1]

*Pittsburgh Water and Sewer Authority, Pittsburgh, PA, USA, 15238

**University of Pittsburgh, Swanson School of Engineering, Pittsburgh, PA 15261

*Email: casson@pitt.edu

ABSTRACT

In 2010, The Pittsburgh Water and Sewer Authority (PWSA), and several other drinking water utilities that draw their source water from the Allegheny River in Western Pennsylvania, observed a dramatic increase in trihalomethane (THM) concentrations in their finished water. The increase was especially pronounced for the brominated species. A couple of these water systems had recently violated the Stage I Disinfectant/Disinfection Byproduct (D/DBP) Rule. Some of these utilities, and others, have expressed concern over their ability to comply with the more stringent requirements of the Stage II D/DBP Rule that goes into effect in 2012. Subsequent analysis of a number of water samples collected from the Allegheny River indicated the presence of elevated concentrations of bromide which could explain the observed changes in the quantity and chemical nature of drinking water THMs.

In response to these observations, PWSA and the University of Pittsburgh, School of Engineering undertook an investigation of river bromide and drinking water THM formation in the Allegheny River basin. As part of this investigation, the bromide content of Allegheny River water at the intake of the PWSA Drinking Water Treatment Plant was measured on a daily basis. These data were subsequently correlated with weekly THM concentrations in finished water leaving the PWSA plant. Laboratory trials were conducted to evaluate THM formation in river water experimentally supplemented with excess bromide. An effort was also made to determine the effectiveness of conventional drinking water treatment processes in removal of bromide from source water.

In parallel with these studies, an extensive survey was conducted measuring the concentration of bromide, total dissolved solids, and sulfates in water samples collected on a monthly basis along the length of the Allegheny River and its major tributaries. Special emphasis was placed on identifying industrial activities that may contribute bromide to the river. These included municipal and industrial wastewater facilities that treat flowback water generated by Marcellus Shale gas drilling operations; treated and untreated acid mine drainage discharged from coal mines, coal fired power plants that may be using brominated compounds to control mercury in air emissions; and coal fired power plants and steel mills that may be using brominated compounds to control biological growth in cooling towers.

Results of this investigation indicated that elevated concentrations of bromide in river source water are associated with increased concentrations of total trihalomethanes, especially the more toxic brominated THMs, in finished drinking water. The results also indicated that conventional drinking water treatment processes are ineffective in removing bromide from the source water. Additionally, the river survey suggested that industrial wastewater treatment plants treating Marcellus Shale flowback water and other wastes appear to be a major contributor of bromide to the Allegheny River. Certain coal fired power plants also appear to intermittently contribute bromide to the river, but to a lesser extent.

The apparent increase in river bromide concentrations in the Allegheny River Basin may significantly affect the ability of drinking water treatment plants to comply with the D/DBP Rule.

KEYWORDS: Drinking Water, THMs, Marcellus Shale Drilling, Disinfection By Products

INTRODUCTION

Rook (1974) and Bellar et al, (1974) first described the formation of THMs during chlorination of surface waters almost 40 years ago. The United States Environmental Protection Agency (EPA) began regulating these substances in American drinking waters when they promulgated the Trihalomethane Rule in 1979. A number of US drinking water utilities are currently struggling to comply with Stage I of EPA's Disinfectant/Disinfection Byproduct (D/DBP) Rule. Even more utilities are anticipating difficulty meeting the more stringent Stage II D/DBP Rule which goes into effect in 2012. Water companies are utilizing a number of approaches to ensure compliance. These include optimization of the treatment process to more effectively remove disinfection byproduct precursors; optimization of the distribution system to minimize the age of treated water; substitution of free chlorine with disinfectants like chloramines, ozone, or UV light that form fewer THMs; and air stripping in finished water storage tanks to remove volatile THMs from drinking water.

Disinfection byproducts are formed when source water containing disinfection byproduct precursors (i.e., Natural Organic Matter such as humic and fulvic acids) and bromide come into contact with chlorine in a drinking water treatment plant and distribution system. If bromide is present in low concentrations in the source water, then the trihalomethanes that are formed consist primarily of chlorinated species (e.g., chloroform). However, if elevated concentrations of bromide are present in the source water, a greater proportion of brominated trihalomethanes are formed (e.g., bromoform, dibromochloromethane, and bromodichloromethane). During the drinking water treatment process, chlorine oxidizes bromide to form hypobromous acid. Hypobromous acid behaves as a stronger, more rapidly reacting substituting agent than hypochlorous acid thereby resulting in the formation of greater amounts of brominated disinfection byproducts than if excess concentrations of hypobromous acid were not present (Singer and Reckhow, 2011). Additionally, the final concentration of Total Trihalomethanes (expressed on a weight to volume basis as ug/L) is greater at equal molarity if excess bromide is present in the source water because bromide is heavier than chloride. The molecular weight of chloroform is 119 gm/mole vs 253 gm/ mole for bromoform. The formation of high concentrations of THMs, especially brominated THMs, in drinking water derived from source waters containing elevated concentrations of bromide has been well documented in the literature (Oliver, 1980; Minear and Bird, 1980; and Luong et al., 1982). A well known case study is the production of high concentrations of brominated THMs in Israel's national water supply resulting from naturally elevated bromide concentrations (2 mg/L) in the source water obtained from Lake Galilee (Rebhun et al, 1988). Brominated THMs typically account for >90% of TTHMs in this drinking water with bromoform being the dominant species.

The primary toxic endpoints of concern from chronic exposure to low levels of DBPs in drinking water are cancer and reproductive/developmental effects. For analogous DBPs, genotoxicity and cytotoxic potency are generally greatest in iodinated compounds, followed by brominated compounds, with chlorinated compounds being the least potent (Richardson et al, 2007).

In September 2010 the Pittsburgh Water and Sewer Authority (PWSA) observed significantly increased concentrations of total trihalomethanes in tap water samples routinely collected from its distribution system. Not only were the total THMs elevated but the percentage concentration of brominated species was substantially higher than normal. In an attempt to determine whether this phenomenon was unique to the Pittsburgh utility, PWSA arranged, with the assistance of the Pennsylvania Department of Environmental Protection, to analyze finished water samples collected from six other drinking water companies located upstream of the PWSA intake on the Allegheny River. As was the case for the Pittsburgh system, the concentrations of total trihalomethanes, as well as the percentage of brominated THMs, were unusually high in the samples from the other utilities. This suggested that the observed change in the quantity and chemical nature of disinfection byproducts may be associated with a change in the chemical characteristics of the river source water, specifically in the concentration of bromide. A similar increase in total THM concentrations, and an increase in brominated species, apparently associated with increased bromide concentrations in the river, had been observed in the nearby Monongahela River two years earlier (Handke, 2009). Although bromide had not previously been analyzed in samples collected from this portion of the Allegheny River, samples collected at this time seemed to contain elevated concentrations, ranging as high as 241 ppb (ug/L) bromide.

The observation of elevated THMs in drinking water and the apparent elevated levels of bromide in the Allegheny River suggested three questions:

- **What effect does excess bromide in river source water have on THM formation in a drinking water system?** Specifically, how do excess bromides affect the concentration of total THMs and the percentage contribution of brominated THMs?

95

- **How effective is conventional drinking water treatment removing bromide from source water?**
- **What is the natural background level of bromide is in the Allegheny River, how much does it vary (geographically and temporally), and what are the sources of excess bromide?** Specifically, is bromide contributed by industrial sources such as coal-fired power plants, steel mills, municipal sewage treatment plants or industrial wastewater treatment plants treating Marcellus Shale flowback water, and/or facilities discharging treated and untreated acid mine drainage from coal mines?

The Pittsburgh Water and Sewer Authority, in conjunction with the University of Pittsburgh School of Engineering, conducted a comprehensive, research project to answer these questions.

MATERIALS AND METHODS

Impact of Source Water Bromide Levels on THM Formation

To address the question concerning the effect of elevated bromide concentrations on production of total trihalomethanes, and specifically brominated THMs, both a treatment plant survey and a series of laboratory experiments were conducted.

In the treatment plant survey, daily river water samples collected from the intake of the PWSA Treatment Plant were analyzed for bromide content and the values were averaged, for Sunday and Monday of each week, over a period of one year. Finished water samples leaving the PWSA Treatment Plant were collected each Wednesday and analyzed via gas chromatography to measure the concentrations of the four trihalomethanes. The percentage brominated THM species was then correlated with river bromide concentration to evaluate the impact of bromide in source water on speciation of THMs leaving the treatment plant. Sunday/Monday river bromide concentrations were correlated with Wednesday finished water THM concentrations because the average detention time in the PWSA Treatment Plant is two to three days. Comparisons were not made between river bromide levels and total THM concentrations due to concerns that the impact of varying source water temperatures over the seasons would mask the relationship between river bromide concentrations and total THM production.

Laboratory experiments were performed to evaluate the influence of source water, supplemented with various concentrations of bromide, on THM formation. Aliquots of water collected from the Allegheny River were supplemented with various concentrations of bromide ranging from 20 to 300 ppb. Chlorine was then added to the spiked subsamples to produce a concentration of approximately 1.5 ppm (mg/L) free chlorine. The subsamples were subsequently placed in a water bath in which the temperature was maintained at 30° C for a period of 14 days. Following the incubation period, THM formation was stopped by the addition of sodium thiosulfate and the subsamples were analyzed for trihalomethane content. The purpose of these trials was to evaluate the impact of increasing concentrations of bromide in the source water on subsequent THM formation in river water disinfected with free chlorine. A temperature of 30° C and a 14-day incubation period were utilized in these trials to simulate THM formation conditions during summer months at the most distant reaches of the PWSA finished water distribution network.

Effectiveness of Conventional Drinking Water Treatment in Removal of Bromide from Source Water

To investigate the effectiveness of conventional drinking water treatment processes in removal of bromide from source water, a series of water samples were collected from various locations in the PWSA Treatment Plant beginning at the river intake and ending at the finished water entry point to the distribution system. These samples were collected over a several-day period at intervals designed to correspond with the plug flow movement of water through the treatment plant and were analyzed for bromide content.

Bromide in Allegheny River Water

In order to determine the background concentrations of bromide in Allegheny River water, temporal and geographic variations, and sources of excess bromide, a comprehensive survey of bromide concentrations in the Allegheny River and its tributaries was conducted on a regular basis for a one year period. Bromide was measured on water samples collected from the PWSA Treatment Plant intake on a daily basis. Additionally, samples were collected from a total of 42 locations along the main stem of the Allegheny River and several of its tributaries on a monthly basis. A number of the sampling sites were selected to assay bromide concentrations upstream and downstream of industrial facilities potentially contributing bromide to the river system. These included six coal-fired power plants, one steel mill, two locations from which treated and untreated acid mine drainage from coal mines discharges into the river, four municipal sewage treatment plants known to have processed flowback water from Marcellus Shale drilling operations, and four industrial wastewater treatment plants known to have treated Marcellus Shale flowback water. Municipal sewage treatment plants are indicated on the maps in this paper as POTWs (publically owned treatment works).

Coal fired power plants and steel mills are potential candidates for bromide discharge because some use brominated compounds to help control biological growth in cooling towers. Coal fired power plants are also a potential source of bromide because some of them utilize brominated compounds to help control mercury in air emissions. Acid mine drainage is a possible source of bromide due to the possible presence of bromide in coal beds.

Municipal sewage treatment plants and industrial wastewater treatment plants treating Marcellus Shale wastewater may also be a source of bromide to the river system. While there are a number of shale deposits throughout the US containing abundant natural gas resources, Marcellus Shale is the most expansive shale gas in play in this country. The Middle Devonian Marcellus Shale formation underlies a large portion of the Appalachian Basin including Pennsylvania and parts of West Virginia, New York, Ohio, and Maryland (Kargbo et al, 2010; Chapman et al, 2012). To extract natural gas from geologic deposits like the Marcellus Shale layer a process termed high-volume horizontal hydraulic fracturing or 'fracking' is used. This process involves injection of 2 to 15 million gallons of water, under high pressure, as deep as one mile into each well, and laterally like spokes of a wheel, for distances of up to one mile from the wellbore, to create fissures in the deep rock layer allowing the release of gas (Kerr, 2010). A variety of chemicals (e.g., acids, surfactants, scale inhibitors, and biocides) along with sand are injected with the water to facilitate the extraction of natural gas. Typically 20% to 30% of the water injected into the well eventually returns to the surface as either flowback water, which returns to the surface within a short time after fracking, or produced water, which comes to the surface over time with the gas. The flowback and produced waters typically have a very different chemical composition than the injected water. Flowback and produced waters have total dissolved solid (TDS) contents exceeding 200,000 mg/L, with elevated levels of strontium, bromide, sodium, calcium, barium, and chloride most likely from interaction with formation waters and salts in the producing formation (Blaunch et al, 2009).

Wastewater from Marcellus Shale operations may be disposed of in several ways. It may be recycled and used for additional fracking operations; it can be returned underground into permitted deep well injection sites, or it can be discharged to surface waters following treatment. Municipal sewage treatment processes and industrial wastewater treatment processes, other than ion exchange or reverse osmosis, are relatively ineffective in removing bromide from flowback water. The commercial wastewater treatment plants on the Allegheny River System had been built years earlier to treat wastewater from conventional gas and oil wells and discharge the treated waste to the river. These commercial plants have more recently been used to treat wastewater from unconventional deep well gas drilling sites and discharge this treated waste into surface waters. Since these plants were not designed to remove the high levels of total dissolved solids and metals contained in wastewater from deep well drilling, it is expected that these treated effluents discharged into rivers may contain high concentrations of bromide.

Bromide analyses were also conducted on two Allegheny River water samples collected quarterly from locations on the main stem of the river believed to be upstream of most industrial contamination. These sites (Tionesta Bridge at river milepoint 152, and Warren Bridge at milepoint 188) are situated far upstream of the PWSA intake (mile point 8.4). These sample locations are important because they are believed to provide baseline data on natural, background bromide concentrations in water from the Allegheny River System.

In addition to bromide measurements, river samples collected on a daily, monthly, or quarterly basis were also analyzed for a number of other chemical parameters including: total Dissolved Solids (TDS) and sulfate concentration.

THM analyses were conducted by gas chromatography (EPA Method 502.2). Bromide was analyzed via ion chromatography (EPA Method 300.1). Total Dissolved solids were measured by gravimetric analysis (Standard Method 2542). And, sulfate was analyzed using a turbidimetric method (Standard Method 4500 SO_4-E).

RESULTS AND DISCUSSION

Impact of Source Water Bromide Levels on THM Formation

A treatment plant survey was conducted to evaluate the impact of bromide concentrations at the PWSA Treatment Plant intake on THM formation 2 to 3 days later in finished water leaving the treatment plant. Figure 1 illustrates the direct relationship between source water bromide concentrations and the percentage of brominated species in finished water THMs. The correlation coefficient is statistically significant at the 0.001 level. As the figure indicates, at a concentration of 50 ppb bromide in the source water, approximately 62% of the finished water THMs consists of bromoform, dibromochloromethane, and bromodichloromethane. However, at 150 ppb bromide in the river water, approximately 83% of finished water THMs consists of brominated species. This is an important increase since brominated trihalomethanes are considered to be more carcinogenic than chloroform. It is also important because brominated THMs have a lower Henry's Law Constant than nonbrominated THMs and are therefore more difficult to remove from distribution system storage tanks for drinking water utilities utilizing air stripping to remove THMs.

Laboratory trials were conducted to assess the impact of raw source water artificially supplemented with increasing concentrations of bromide on total THM formation (Table 1). As Figure 2 shows, the concentration of total THMs increased from 105 ppb at a source water bromide concentration of 50 ppb to 120 ppb at a bromide concentration of 200 ppb. The percentage concentration of brominated THMs increased from approximately 25% at 50 ppb bromide to 70% at 200 ppb bromide.

Effectiveness of Conventional Drinking Water Treatment Processes in Removal of Bromide from Source Water

An important question in this study is the effectiveness of conventional drinking water treatment processes, such as those utilized at the Pittsburgh Water and Sewer Authority Treatment Plant, in removing bromide from source water. Do the processes of non chlorine-pre oxidation, carbon adsorption, clarification, and filtration make bromide unavailable for formation of brominated disinfection byproducts before the chlorine disinfection step is reached in the treatment chain?

The treatment process utilized at the Pittsburgh Drinking Water treatment Plant is similar to that used in many treatment plants around the world. Approximately 70 million gallons of water from the Allegheny River is pumped into the plant each day. Potassium permanganate is fed, for oxidation of dissolved iron and manganese, as the water passes through the intake. About 15 minutes later ferric chloride (25 mg/L) and a cationic polymer (1 mg/L) are added for coagulation, and lime is added for adjustment of pH to a level of 6.7. This concentration of coagulant and this particular pH level are utilized to facilitate the 'Enhanced Coagulation' process to optimize removal of THM precursors. Powdered activated carbon (1 mg/L) is added for adsorption of disinfection byproduct precursors. The water then passes through a three stage flocculation process and is allowed to settle for a two hour period in the clarifier. The water then settles for an additional period of 24 hr. in a 117 million gallon capacity secondary sedimentation basin to complete the clarification step. Following this, a small amount of sodium hypochlorite is dosed into the water, just prior to filtration, to oxidize any remaining dissolved manganese to ensure removal during the filtration process. Rapid sand filtration involves passage of the water through 18 in. of coal, 12 in. of sand, and 12 in. of support gravel. Sodium hypochlorite is then added for disinfection, hydrofluorosilic acid for fluoridation, and sodium carbonate for final pH adjustment. The water subsequently passes through a 44 million gal. capacity clearwell, allowing plenty of time for disinfection, before it leaves the treatment plant and enters the finished water distribution system. The entire treatment process occurs over a three day period.

Two surveys were conducted, to measure bromide removal, as water passes through the various steps of the treatment process. In one survey the initial river concentration of bromide was quite high (188 ppb), while in the second survey the initial river concentration was much lower (44 ppb). As the results in Table 2 indicate, a small amount of bromide was removed, in both trials, during primary clarification. This removal is likely due to oxidation by potassium permanganate. Potassium permanganate oxidizes bromide to hypobromous acid which is quite reactive and is probably consumed by oxidation of reduced chemical substances naturally present in the source water.

Following the initial, partial removal of bromide, the next reduction occurs when the pre-filtered water passes through the treatment plant rapid sand filters, and when the post-filtered water passes through the clearwell and becomes finished water. It is at this point that the major dose of free chlorine is added to the treatment process at the PWSA Plant. As the table indicates, it is during these final treatment stages that bromide is removed to concentrations lower than the limits detectable by ion chromatography. At this point free chlorine oxidizes bromide to hypobromous acid which then reacts with THM precursors (e.g., humic and fulvic acids) to form brominated trihalomethanes.

As these trials indicate, the only significant removal of bromide during the conventional drinking water treatment process is the oxidation of bromide by free chlorine which leads to formation of brominated THMs. Therefore, it appears that conventional drinking water treatment processes are generally ineffective in removing bromide so that it is not available for subsequent formation of disinfection byproducts.

Bromide in the Allegheny River

Important questions addressed by the current study include: how much bromide is in the Allegheny River (especially at the PWSA Treatment Plant intake); how and why do bromide concentrations in the Allegheny River vary geographically and temporally; and which industrial or natural processes along the Allegheny and its tributaries contribute bromide to the river system?

Table 3 summarizes daily concentrations of bromide in the PWSA raw water intake on the Allegheny River during late 2010 and 2011. As the table indicates, daily bromide concentrations vary significantly from a low of less than 25 ppb to a high of 299 ppb. The results in this table have been color coded with values less than 100 ppb appearing white, values between 100 and 200 ppb appearing yellow, and values greater than 200 ppb shaded red. Assuming that bromide contributions to the river system from natural or industrial sources are relatively constant throughout the year, one explanation for the wide fluctuation in bromide values at the PWSA raw water intake may be river flow or the volume of water in the river. Figure 3 compares daily river flow, measured at the closest US Geological Survey stream gauging station to the PWSA Plant (i.e., Natrona, PA) with bromide concentrations at the PWSA intake. As the figure shows,

100

bromide concentrations tend to be greatest when river flow is lowest. This suggests that regardless of the source of bromide in the Allegheny system, drinking water utilities such as the Pittsburgh Water and Sewer Authority are exposed to the greatest challenge from bromide during low river flow conditions.

In an attempt to identify potential sources of excess bromide in the Allegheny River and its tributaries, an extensive survey of the river system was performed. Following some initial testing (September through December, 2010), a number of locations along the river and its tributaries were sampled on a monthly basis throughout 2011 and analyzed for bromide content, total dissolved solids, and sulfate concentration. Figure 4 is a map of the Allegheny River and its major tributaries. The PWSA intake is located at the base of the Allegheny River main stem, just 8.4 miles upstream of the point where the Allegheny converges with the Monongahela River to form the Ohio River. A special emphasis was placed on analyzing monthly water samples collected from points upstream and downstream of industrial sites suspected of contributing excess bromide to the river. These included six coal fired power plants; one steel mill; two areas on the river that are known to contain treated and untreated acid mine drainage; as well as four municipal wastewater treatment plants and four industrial wastewater treatment plants that are known to have treated wastewater generated by Marcellus Shale drilling operations.

Table 4 summarizes bromide data from samples collected on a routine basis along that stretch of river. As the table indicates, bromide concentrations in river water samples collected from the towns of Tionesta and Warren, PA are very low, typically less than 50 ppb. These towns are located far upstream (river mile points 152 and 188, respectively) and consequently were only sampled on a quarterly basis during the project. It is believed that there are no major industrial discharges of bromide occurring upstream of these two towns. Therefore, it may be concluded that these low bromide concentrations represent natural, background levels of bromide for the Allegheny River. A substantial increase in river bromide concentration appears in the downstream vs. upstream comparison surrounding Industrial Wastewater Treatment Plant A. This plant is known to treat flowback water from Marcellus Shale drilling operations. Since this plant is discharging effluent directly into the main stem of the Allegheny River, where there is a significant dilution factor, even 2X, 3X, and 4X increases in bromide concentrations in the river at this point probably represent significant contributions of bromide from this plant. No notable increase in river bromide levels appears to be associated with coal fired power plant E.

As the data in Table 5 indicates, there is no apparent contribution of bromide from POTW A. Interestingly, during most of the months from May through October, there is an unusually high concentration of bromide measured at the Ridgeway Bridge. Since there are no known waste treatment sites upstream of POTW A, these data may suggest the possibility of illegal discharge of untreated wastewater into the upper reaches of the Clarion River. Discharges of this type, involving drilling wastewater, have been reported previously in other parts of the Allegheny River System and have been prosecuted by the Pennsylvania Department of Environmental Protection (PADEP). It is interesting to note that the upstream increases of bromide coincided with the May 2011 implementation of a voluntary ban, requested by the Pennsylvania Department of Environmental Protection (PADEP), on discharge of treated Marcellus Shale wastewater into the state's surface water systems. This ban is discussed in greater detail later in this paper.

Again, the relatively low bromide concentrations measured in samples collected from the 'Bridge Upstream of Industrial Waste Plant B' and from 'Blue Spruce Bridge' presumably represent the low natural background levels of bromide in the Allegheny River and its tributaries (Table 6). The dramatic increase in bromide concentrations (extending up to a 34 fold increase) in samples collected from 'Bridge Street Bridge' suggests a significant introduction of bromide into the tributary associated with Industrial Waste Plant B. Downstream bromide concentrations in several of the sampling months (July and August 2011) exceed 3,000 ppb (3 ppm). As samples collected from 'Stitt Hill Road Bridge' indicate, bromide levels become naturally diluted as they pass through Crooked Creek Lake prior to entering the main stem of the Allegheny River. However, the 'Stitt Hill Road Bridge' bromide concentrations are still substantially greater than those in samples collected near the upstream ends of the tributaries (i.e., 'Bridge Upstream of Industrial Waste Plant B' and 'Blue Spruce Bridge') suggesting a net contribution of bromide to the Allegheny main stem associated with Industrial Waste Plant B.

As Table 7 shows, the concentrations of bromide in water samples collected at sites upstream of known industrial discharges ('Neal Road Bridge', 'Route 56 Bridge in Armaugh', and 'Johnstown Railroad Bridge') are almost always less than 100 ppb, and usually less than 50 ppb. These values typify the low, natural concentrations of bromide in the river system. Increases in bromide concentration ranging up to 21 fold downstream vs. upstream of Industrial Waste Plants C suggests significant discharges of bromide to natural waters from this facility. Increases in bromide levels, ranging up to 7X downstream vs. upstream of POTW B, suggest sporadic releases of excess bromide from this municipal wastewater treatment plant which has been known to process Marcellus Shale wastewater. Increases in bromide concentrations, reaching 9X in downstream vs. upstream measurements of coal fired Power Plants A and B, suggest periodic releases of bromide from these facilities as well. The apparent power plant discharges seem to be greater during the summer months. This could be associated with seasonal use of brominated anti-fouling compounds in the cooling towers of these plants during the warmer seasons of the year.

As Table 8 indicates, the low concentrations of bromide in Loyalhanna Creek (almost always less than 50 ppb), a tributary flowing into the Kiski River, again demonstrates the low natural background levels of bromide in the Allegheny River System. This contrasts sharply with the elevated concentrations of bromide in the upper portion of the Kiski River (e.g., 'Washington Street Bridge' sampling site) which is heavily influenced by industrial discharges of bromide on the Conemaugh River and its tributaries. The absence of an apparent increase in bromide concentrations downstream vs. upstream of POTW C is typical of most of the POTWs surveyed in this investigation.

As Table 9 indicates, potential discharge sites for bromide in this stretch of the river do not suggest significant contributions of bromide to the river. This includes Steel Plant A, Coal Fired Power Plants C & D, and POTW D. This also includes two locations on the river system, labeled 'Buffalo Creek' and 'Harmar Mine', where there is ongoing discharge of treated and untreated acid mine drainage, respectively. The absence of consistent increases in bromide concentrations at the downstream vs. the upstream sampling locations at these two sites suggests no apparent bromide contribution from acid mine drainage. These observations are consistent with Cravotta's (2008) study in which he measured an average bromide concentration of only 45 ppb in 99 samples of untreated effluent from abandoned bituminous coal mines in Pennsylvania.

45 ppb is approximately the background concentration of bromide observed in our survey of the upper reaches of the Allegheny River and its tributaries.

Mass Balance Analysis of Bromide in the Allegheny River System

Mass balances were calculated using stream flow data (Table 10). United States Geological Survey stream flow data were available for river or stream locations near several of the industrial facilities that were shown in the year-long survey to be discharging bromide. These facilities included Industrial Waste Plants A, C, and D, as well as Coal Fired Power Plants A&B. The mass of bromide (expressed in terms of lb/day) was calculated for monthly samples collected upstream and downstream of these apparent discharge sites using measured bromide concentration data and flow data. Because flow data were not available for either McKee Run or Crooked Creek in the vicinity of Industrial Waste Plant B, the bromide mass contribution from this site could not be calculated.

A substantial portion of the increase in bromide as water flows down the Allegheny River System could be accounted for by the other three industrial wastewater treatment plants and Coal Fired Power Plants A&B. For example, in August 2011 the difference between the single day bromide mass calculated at Franklin Bridge (1,454 lb/day) and the mean daily bromide mass calculated at the PWSA Treatment Plant intake (4,632 lb/day) was 3,178 lb/day. The sum of individual contributions of bromide mass from Industrial Waste Plants A, C, & D, and from Power Plants A&B, totaled 1,622 lb/day of bromide. These data suggest that the industrial wastewater sites may have accounted for approximately 51% (1,622 lb per day/3,178 lb per day) of the increase in bromide mass as water flowed down the Allegheny River. The mean monthly percentage contribution, over 12 months, of the various industrial sites to the typical increase in bromide mass between the upstream Franklin Bridge sampling location and the downstream PWSA intake is 52%. While this mass balance approach is only a rough approximation, due to the necessity of using a monthly average for bromide mass at the PWSA intake, the estimate suggests that the industrial sites are significant contributors of excess bromide to the Allegheny River System.

Regulatory Control of Discharge of Drilling Wastes into Natural Bodies of Water

In April 2011, the Pennsylvania Department of Environmental Protection requested drillers to voluntarily stop using municipal and industrial wastewater treatment plants for treatment and ultimate disposal of Marcellus Shale wastewater to surface waters. However, this survey indicates a continued discharge of bromide from several industrial locations after that request was made.

In October 2011, the US Environmental Protection Agency announced plans to develop national standards for treatment of wastewater discharges produced by natural gas extraction from underground coal bed and shale formations (US Environmental Protection Agency, 2011). EPA will consider standards based on demonstrated, economically achievable technologies. After gathering data, and consulting with stakeholders, EPA anticipates soliciting public comment on a proposed rule for coal bed methane in 2013 and for shale gas in 2014.

Future Research

The survey of bromide in the Allegheny River and its tributaries suggests several sites as contributors of bromide to the river system (e.g., industrial wastewater treatment plants, certain coal-fired power plants). Because industrial wastewater treatment plants may treat a variety of industrial wastes (including Marcellus Shale drilling flowback and produced waters, coal mine drainage, brine from abandoned oil and gas wells from formations shallower than Marcellus, etc.), there is some uncertainty as to the ultimate source of the excess bromide being discharged from these facilities. The next phase of this research will involve an additional collaborative effort with the University of Pittsburgh, Department of Geology to identify the source of the bromide laden total dissolved solids. Chapman et al, (2012) described a technique whereby they can differentiate between Marcellus Formation produced water and other potential sources of TDS in ground and surface waters by analyzing strontium isotope ratios ($^{87}Sr/^{86}Sr$)

SUMMARY and CONCLUSIONS

In late 2010, the Pittsburgh Water and Sewer Authority observed a dramatic increase in trihalomethane concentrations, and the percentage of brominated THMs, in drinking water samples collected from its distribution system. Figure 5 summarizes THM compliance data for a seven-year period based on Stage I of the Disinfectants/Disinfection Byproduct Rule. The percentage of brominated THMs observed in 2010 (88%) was significantly higher than for any of the other years. Additionally, despite continued efforts to reduce the concentration of total trihalomethanes by optimizing the treatment and distribution processes, the mean concentration of TTHMs for 2010 was the second highest for the seven year period. Because of concerns over these observations and over PWSA's ability to meet the more stringent requirements of the newly promulgated Stage II Disinfectants/Disinfection Byproduct Rule, the water utility and the University of Pittsburgh initiated a comprehensive study of bromide in the Allegheny River and its impact on disinfection byproduct formation.

The results obtained in this study suggest the following conclusions:

- Increased bromide concentrations in river source water lead to elevated total trihalomethane concentrations, and an increase in the percentage contribution of the more toxic brominated THMs, in finished drinking water.
- Conventional drinking water treatment processes do not effectively remove bromide from source water.

- Radionuclides do not appear to be elevated in the Allegheny River or its tributaries, despite the apparent discharge of treated and possibly untreated Marcellus Shale drilling wastewater into the river system.
- Bromide concentrations appear to be naturally low in the Allegheny River system as indicated by the consistently low levels measured in water from upstream sites on the Allegheny River and its tributaries. Bromide concentrations tend to increase as water moves downstream in the Allegheny System. However, all of the increase appears to be associated with point discharges of bromide from industrial sites rather than with the accumulation of bromide from the natural environment. Following the introduction of bromide at an industrial site, bromide levels tend to decrease naturally with dilution as the water continues to flow downstream.
- Bromide concentrations are significantly affected by river volume (e.g., snow melt, heavy rains). Therefore, potential THM problems in drinking water, associated with excess river bromide, are more acute during low river flow conditions.
- Total Dissolved Solids is not a reliable indicator or surrogate parameter for bromide in the Allegheny River.
- Bromide concentrations do not appear to increase downstream of most municipal wastewater treatment plants treating Marcellus Shale wastewater. A possible explanation may be that most municipal sewage plants chlorinate their treated effluent for pathogen control prior to discharge to surface waters. Bromide in the chlorinated effluent is likely oxidized to hypobromous acid which is then consumed by reducing compounds in the sewage plant effluent. Alternately, brominated THMs may be formed at this point and would be subsequently volatilized to the atmosphere as the river water flows downstream.
- Bromide concentrations increase downstream of industrial wastewater treatment plants treating Marcellus Shale wastewater. This suggests that the treatment processes in these plants are ineffective in removing bromide.
- Bromide concentrations occasionally increase downstream of some of the coal fired power plants. However, the discharge of bromide appears to be less than that observed at industrial wastewater treatment plants.
- Bromide concentrations do not appear to increase downstream of the single steel plant or the two acid mine drainage sites surveyed in this study.
- The concentration of bromide from industrial wastewater treatment plants and certain coal-fired power plants may account for approximately 50% of the increase in bromide as water flows through the Allegheny River System.

REFERENCES

Bellar, T.A.; Lichtenberg, J.J.; & Kroner, R.C., 1974. The Occurrence of Organohalides in Chlorinated Drinking Water. *Jour. AWWA*, 66:12:703.

Blaunch, M.E.; Myers, R.R.; Moore, T.R.; & Lipinski, B.A., 2009. Shale Post-Frac Flowback Waters – Where Is All the Salt Coming From and What Are the Implications? 2009 Society of Petroleum Engineers Eastern Regional Meeting, SPE 125740, Charleston, WV, 2009; pp 1-20.

Chapman, E.C.; Capo, R.C.; Stewart, B.W.; Kirby, C.S.; Hammack, R.W.; Schroeder, K.T.; & Edenborn, H.M., 2012. Geochemical and Strontium Isotope Characterization of Produced Waters from Marcellus Shale Natural Gas Extraction. *Environ. Sci. & Technol.* 46:3545.

Cravotta III, C.A. 2008. Dissolved Metals and Associated Constituents in Abandoned Coal Mine Discharges, Pennsylvania USA Part 1: Constituent Quantities and Correlations. *Applied Geochem.*, 23:166.

Handke, P. 2009. Trihalomethane Speciation and the Relationship to Elevated Total Dissolved Solids Concentrations Affecting Drinking Water Quality at Systems Utilizing the Monongahela River as a Primary Source During the 3rd and 4th Quarters of 2008. Pennsylvania Department of Environmental Protection. pp1-18.

Hill, D.G.; Lombardi, T.E.; & Martin, J.P., 2004. Fractured Shale Gas Potential in New York. *Northeastern Geol. Environ. Sci.* 26:8.

Karkbo, D.M.; Wilhelm,R.G.; & Campbell, D.J., 2010. Natural Gas Plays in the Marcellus Shale: Challenges and Potential Opportunities. *Environ. Sci. & Technol.* 44:15:5879.

Kerr, R.A., 2010. Natural Gas from Shale Bursts onto the Scene. *Science.* 328:1624.

Luong, T.V.; Peters, C.J.; & Perry, R., 1982. Influence of Bromide and Ammonia Upon the Formation of Trihalomethanes Under Water Treatment Conditions. *Environ. Sci. & Technol.* 16:476.

Minear, R.A. & Bird, J.C., 1980. Impact of Bromide Ion Concentration on Yield, Species Distribution, Rate of Formation and Influence of Other Variables. *Water Chlorination: Environmental Impact and Health Effects.* Vol. 3 (R.L. Jolley et al, editors). Ann Arbor Sci. Publ., Ann Arbor, Mich.

New York State Department of Health, 2009. Marcellus Shale Potential Public Health Concerns. http://s3.amazonaws.com/propublica/assets/natural_gas/nysdoh_marcellus_concerns_090 721.pdf

New York Times, 2011. Regulation Lax as Gas Wells' Tainted Water Hits Rivers. February 26, 2011.

Oliver, B.G., 1980. Effect of Temperature, pH and Bromide Concentration on the Trihalomethane Reaction of Chlorine with Aquatic Humic Material. *Water Chlorination: Environmental Impact and Health Effects,* Vol. 3 (R.L. Jolley et al, editors). Ann Arbor Sci. Publ., Ann Arbor, Mich.

Rebhun, M.; Manka, J.; & Zilberman, A., 1988. Trihalomethane Formation in High Bromide Lake Galilee Water. *Jour. AWWA.* 80:6:84.

Richardson, S.D.; Plewa, M.J.; Wagner, E.D.; Schoeny, R.; & Demarini, D.M., 2007. Occurrence, Genotoxicity, and Carcinogenicity of Regulated and Emerging Disinfection By-products in Drinking Water: a Review and Roadmap for Research. *Mutat. Res.* 636: 178-242.

Rook, J.J., 1974. Formation of Haloforms During Chlorination of Natural Water. *Jour Water Treatment & Examination.,* 23:3:234.

Singer, P.C. & Reckhow, D.A., 2011. Chemical Oxidation. Chapt 7 in: *Water Quality and Treatment* 6th ed. American Waterworks Association.

US Environmental Protection Agency, 2011. EPA announces schedule to develop natural gas wastewater standards. EPA Press Release, Oct. 20, 2011.

US Environmental Protection Agency, 2012. Safe Drinking Water Act. http://water.epa.gov/lawsregs/rulesregs/sdwa/index.cfm

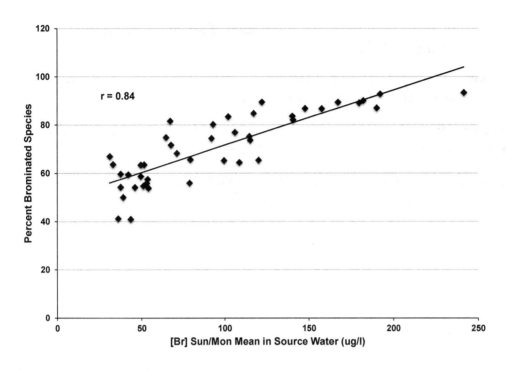

Figure 1 Correlation between Percent Brominated THMs in Finished Water Leaving the Pittsburgh Water and Sewer Authority Treatment Plant and Bromide Concentration in the Allegheny River Source Water.

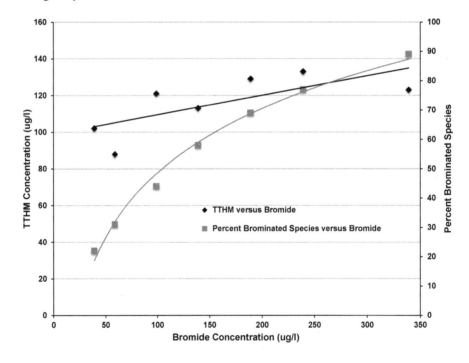

Figure 2 TTHM Formation Potential Study – Effect of Experimental Addition of Bromide.

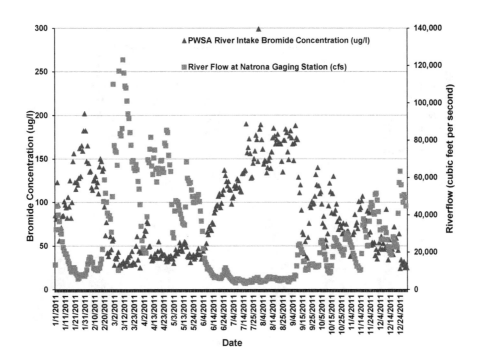

Figure 3 Bromide Concentration and River Flow.

Figure 4 The Allegheny River System.

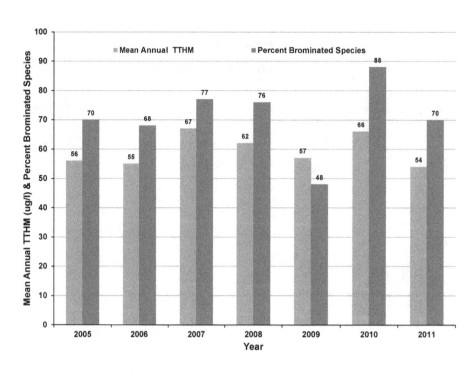

Figure 5 Mean Annual TTHM Concentrations and Percent Brominated Species.

Table 1
TTHM Formation Potential
(Effect of Experimental Addition of Bromide)

Bromide Supplement (ug/l)	Total THMs (ug/l)	Percent Concentration of Bromoform	Percent Concentration of Brominated Species
0*	102	1	22
20	88	1	31
60	121	1	44
100	113	3	58
150	129	5	69
200	133	10	77
300	123	27	89

Notes:

*Baseline Bromide Concentration = 39 ug/l

110

Table 2
Removal of Bromide
by the
Pittsburgh Water and Sewer Authority Drinking Water Treatment Plant

SAMPLE SITE	2010		2011	
	Date – Time	Bromide Concentration (ug/l)	Date – Time	Bromide Concentration (ug/l)
River Intake	25 Oct - 0730	188	21 Mar- 0720	44
Flume	25 Oct – 1200	158	21 Mar-1230	40
Settled Water	26 Oct - 1210	171	22 Mar- 1300	45
Pre-filtered Water	26 Oct - 1515	192	22 Mar- 1600	<25
Post-filtered Water	26 Oct - 1505	134	22 Mar- 1605	<25
Finished Water	27 Oct - 0800	<50	23 Mar- 0800	<25

Table 3
Bromide Concentration (ug/l) at the PWSA Treatment Plant Intake (Allegheny River)

Day of the Month	Sept 2010	Oct 2010	Nov 2010	Dec 2010	Jan 2011	Feb 2011	March 2011	April 2011	May 2011	June 2011	July 2011	Aug 2011	Sept 2011	Oct 2011	Nov 2011	Dec 2011
1			136	37	85	182	58	48	37	35	114	184	176	140	67	35
2			241	42	81	147	28	62	35	37	98	189	182	119	59	42
3			227	39	123	165	36	68	30		105	172	174	110	38	54
4			195	38	97	145	35	67	39	44	107	161	167	112	38	47
5			216	59	56	135	38	78	46	48	116	144	148	98	38	61
6			172	44	66	136	43	76	43	48	119	148		76	63	46
7			230	48	71	117	28	49	41	53	119	147	168	84	66	53
8			170	49	84	114	<25	50	45	66	127	134	188	72	86	46
9			194	53	85	125	29	38	54	63	125	140	174	57	74	48
10			124	58	101	126	30		51	72	117	148	172	50	77	92
11			168	64	97	130	32	33	47	70	113	152	129	64	70	50
12		205	160	68	94	118	30	34	44	68	112	139	111	96	101	34
13		203		49	82	123	27	36	52	77	133	158	120	122	92	57
14		188		57	106	110	32	41	55	75	134	174	91	112	94	48
15		151	170	65	95	141	34	36	54	88	136	185	66	88	102	41
16			155	57	125	150	37	37	49	92	137	176	63	110	107	46
17			165	76	82	147	28	44	38	96		167	41	130	77	40
18			143	35	100	136		55		97	190	172	55	87	72	72
19			146	67	147	139	39	36	37	107	157	178	37	91	79	56
20			158		156	95	28	45	41	110	163	169	36	98	79	59
21			176	88	123	62	44	40	31	94	173	167	60	78	90	52
22			140		115	77	31	39	37	99	156	167	76	69	63	45
23	220		224		124	42	30	40	38	109	128		96	82	88	64
24			204	79	120	38	29	40	40	114	129	147	94	75	66	24
25		188	180	66	128	43	50	35	35	137	145	135	101	71	51	34
26		142	139	106	162	61	61	32	37	131	150	171	98	64	54	30
27		156	145	89	130	46	34	33	40	124	159	171	99	76	52	28
28		190	117	101	165	56	44	38	42	119	177	184	107	66	54	32
29		241	97		159		<25	35	47	124	172	180	132	64	39	25
30		211	79	198	182		42	39	40	115	147	158	120	75	46	29
31		220		98	202		47		40		299	149		67		25
Monthly Rainfall (cm)	8.31	5.38	15.16	3.96	6.12	12.60	12.67	13.03	11.63	6.35	6.65	6.83	9.47	11.18	9.53	6.27
Average Monthly Rainfall (cm)	8.15	5.72	7.67	7.26	6.86	6.02	8.05	7.65	9.65	10.46	10.06	8.84	8.15	5.72	7.67	7.26

Note: Record rainfalls occurred on Nov 25 & 29, 2010

Table 4
Upper Allegheny River
Bromide Concentrations (ug/l)

Sample Site	Sept 2010	Oct 2010	Nov 2010	Dec 2010	Jan 2011	Feb 2011	March 2011	April 2011	May 2011	June 2011	July 2011	Aug 2011	Sept 2011	Oct 2011	Nov 2011	Dec 2011
Warren Bridge				<50 (17th)			<25 (7th)			35 (13th)			33 (30th)			29 (22nd)
Tionesta Bridge				52 (17th)			<25 (7th)			29 (13th)			42 (30th)			40 (22nd)
Franklin Bridge			85 (19th)		63 (21st)	38 (16th)	<25 (30th)	24 (21st)	21 (16th)	57 (23rd)	50 (12th)	94 (2nd)	89 (23rd)	39(13th)	38 (23rd)	92 (23rd)
Industrial Waste Plant A					2 X	3 X	2 X	1.8 X		1.6 X	3 X	1.1 X		4X	1.2 X	
Kennerdell Bridge			83 (19th)		125 (21st)	101 (16th)	51 (30th)	43 (23rd)	20 (16th)	94 (23rd)	134 (12th)	106 (2nd)	56 (23rd)	141 (13th)	45 (23rd)	33 (23rd)
Clarion River																
Armstrong Railroad bridge										87 (23rd)	77 (12th)	89 (2nd)	107 (23rd)	93(13th)	43 (23rd)	24 (23rd)
Coal Fired Power Plant E																
Lock and Dam #8 (RDB)										84 (23rd)	68 (12th)	93 (2nd)	111 (23rd)	95(13th)	67 (23rd)	40 (23rd)
Kittanning Bridge	150 (24th)**	104 (13th) *		50 (28th)	68 (21st)	118 (16th)	30 (30th)	31 (23rd)	26 (16th)	105 (23rd)	82 (14th)	104 (2nd)	121 (23rd)	85(13th)	51 (23rd)	31 (23rd)
Ford City Bridge		101 (13th)**		51 (28th)	57 (12th)	129 (16th)	28 (29th)	39 (13th)	48 (12th)	84 (14th)	82 (14th)	102 (1st)	105 (6th)		59 (8th)	60 (20th)
Crooked Creek	1.1 X	1.1 X			1.1 X	1.2 X	1.2 X				1.4 X	1.1 X	1.4 X	1.1X	2 X	
Schenley (LDB)	170 (24th)	114 (15th)			64 (12th)	146 (16th)	35 (29th)	28 (13th)	47 (12th)	72 (14th)	114 (14th)	114 (1st)	152 (6th)	94(6th)	<25 (8th)	62 (20th)

Notes:

*Mean of RDB and LDB Samples

Table 5
Clarion River
Bromide Concentration (ug/l)

Sample Site	March 2011	April 2011	May 2011	June 2011	July 2011	Aug 2011	Sept 2011	Oct 2011	Nov 2011	Dec 2011
Johnsonburg (Rte. 219 Bridge)									<25 (30th)	25 (30th)
Ridgeway Bridge	<25 (15th)	11 (25th)	171 (31st)		121 (13th)	149 (19th)	44 (29th)	261 (18th)	<25 (30th)	24 (30th)
POTW A		1.3 X			1.1 X		1.6 X	1.1 X		1.9 X
Cherry Tree Trail (LDB)	27 (15th)	14 (25th)	163 (31st)		132 (13th)	156 (19th)	69 (29th)	279 (18th)	26 (30th)	46 (30th)
State Rte. 58 Bridge	<25 (15th)	23 (25th)	37 (31st)		74 (13th)	115 (19th)	105 (29th)	67 (18th)	33 (30th)	16 (30th)

Notes:

LDB = Left Descending Bank

Table 6
Crooked Creek and McKee Run
Bromide Concentration (ug/l)

	Sample Site	Oct 2010	Nov 2010	Dec 2010	Jan 2011	Feb 2011	March 2011	April 2011	May 2011	June 2011	July 2011	Aug 2011	Sept 2011	Oct 2011	Nov 2011	Dec 2011
McKee Run	Bridge upstream of Industrial Waste Plant B		29 (29th)	79 (14th)	61 (12th)	36 (17th)	27 (29th)	17 (13th)	32 (12th)	66 (14th)	93 (14th)	63 (1st)	26 (6th)	25(6th)	44 (8th)	36 (20th)
	Industrial Waste Plant B															
Crooked Creek	Blue Spruce Bridge	57 (29th)	39 (29th)	64 (14th)	87 (12th)	37(17th)	38 (29th)	13 (13th)	53 (12th)	83 (14th)	116 (14th)	114 (1st)	56 (6th)	25(6th)	56 (8th)	52 (20th)
	McKee Run	20 X	10 X	9 X	10 X	1.1 X	3 X	3 X	8 X	8 X	27 X	34 X	4 X	17X	4 X	5 X
Crooked Creek	Bridge St. Bridge	1130 (29th)	345 (29th)	639 (14th)	774 (12th)	42 (17th)	111(29th)	44 (13th)	414 (12th)	640 (14th)	3100 (14th)	3900 (1st)	214 (6th)	427(6th)	244 (8th)	235 (20th)
	Stitt Hill Rd. Bridge		280 (28th)	467 (28th)	396 (12th)	173 (17th)	74 (29th)	53 (13th)	112 (12th)	258 (14th)	578 (14th)	426 (6th)	582 (1st)	103(6th)	116(8th)	167(20th)

Table 7
Conemaugh River and Blacklick Creek
Bromide Concentration (ug/l)

	Sample Site	Oct 2010	Dec 2010	Jan 2011	Feb 2011	March 2011	April 2011	May 2011	June 2011	July 2011	Aug 2011	Sept 2011	Oct 2011	Nov 2011	Dec 2011
Two Lick Creek	Neal Rd. Bridge									90 (28th)	46 (15th)	33 (8th)	41 (10th)	42 (4th)	29 (28th)
	Coal Fired Power Plant F									1.8 X	1.7 X			3 X	
	Hoodlebug Trail (LDB)								151 (30th)	166 (28th)	77 (15th)	31 (8th)	45 (10th)	114 (4th)	27 (28th)
Blacklick Creek	Route 56 Bridge (Armagh)	46 (29th) *		86 (28th)	<25 (24th)	28 (17th)	18 (6th)	45 (17th)	75 (30th)	116 (28th)	50 (15th)	33 (8th)	40 (10th)	38 (4th)	19 (28th)
	Industrial Waste Plant C			11 X	8 X	4 X	5 X	6 X	21 X	21 X	1.7 X	4 X	4 X	13 X	1.3 X
	Rt. 119 Bridge									2400 (28th)	84 (15th)	133 (8th)	39 (10th)	492 (4th)	24 (28th)
	Two Lick Creek										2 X		4 X	2 X	2 X
	Newport Rd. Bridge	<50 (25th)	94 (29th)	961 (28th)	203 (24th)	115 (17th)	87 (6th)	252 (17th)	1600 (30th)	1910 (28th)	210 (15th)	47 (8th)	144 (10th)	290 (4th)	49 (28th)
	Johnstown Railroad Bridge			52 (28th)	<25 (24th)	<25 (17th)	13 (6th)	20 (17th)	32 (30th)	43 (28th)	<25 (15th)	27 (8th)	35 (10th)	<25 (4th)	23 (28th)
	POTW B						1.4 X	1.4 X	1.2 X	1.3 X	7 X				
	Route 56 Bridge (Johnstown)	<50 (25th)	52 (29th)	57 (28th)	<25 (24th)	<25 (17th)	18 (6th)	29 (17th)	39 (30th)	54 (28th)	169 (15th)	<25 (8th)	<25 (10th)	<25 (4th)	14 (28th)
	Coal Fired Power Plants A & B		2 X							4 X	6 X	9 X			2 X
Conemaugh River	Seward Bridge	<50 (25th)	115 (29th)	60 (28th)	<25 (24th)	<25 (17th)	20 (6th)			234 (28th)	1010 (15th)	230 (8th)	<25 (10th)	<25 (4th)	34 (28th)
	Blacklick Creek		4 X		3 X	2 X	3 X			3 X			5 X	4 X	
	Conemaugh Dam Trail (RDB)				82 (24th)	48 (17th)	63 (6th)	146 (17th)	282 (30th)	637 (28th)	342 (15th)	80 (8th)	119 (10th)	101 (4th)	80 (28th)
	Industrial Waste Plant D					1.2 X	1.2 X	1.5 X	1.2 X	1.1 X	1.3 X	1.2 X	1.1 X	1.1 X	
	Tunnelton Rd. Bridge		431 (29th)	237 (28th)	77 (24th)	60 (17th)	78 (6th)	219 (17th)	336 (30th)	711 (28th)	442 (15th)	92 (8th)	131 (10th)	107 (4th)	56 (28th)

Notes:

* Route 259 Bridge

RDB = Right Descending Bank

LDB = Left Descending Bank

Table 8
Kiskiminetas River
Bromide Concentration (ug/l)

Sample Site	Sept 2010	Oct 2010	Nov 2010	Dec 2010	Jan 2011	Feb 2011	March 2011	April 2011	May 2011	June 2011	July 2011	Aug 2011	Sept 2011	Oct 2011	Nov 2011	Dec 2011
Loyalhanna Creek Bridge			33 (24th)	<50 (23rd)	51 (5th)	45 (4th)	<25 (25th)	14 (14th)	18 (18th)	35 (21st)	54 (29th)	43 (16th)	<25 (28th)	30 (11th)	26 (22nd)	30 (20th)
Washington St. Bridge		<50 (25th)	460 (24th)	470 (23rd)	179 (5th)	141 (4th)	70 (25th)	55 (14th)	159 (18th)	434 (21st)	544 (29th)	312 (16th)	145 (28th)	95 (11th)	135 (22nd)	80 (20th)
Avonmore Bridge			390 (24th)	376 (23rd)	147 (5th)	145 (4th)	65 (25th)	54 (14th)	141 (18th)	366 (21st)	315 (29th)	230 (16th)	128 (28th)	95 (11th)	136 (22nd)	87 (20th)
Edmon Bridge			380 (24th)	358 (23rd)	137 (5th)		62 (25th)	57 (14th)	117 (18th)	366 (21st)	304 (29th)	264 (16th)	124 (28th)	96 (11th)	125 (22nd)	84 (20th)
Apollo Bridge			160 (24th)	234 (23rd)	144 (5th)	125 (4th)	70 (25th)	52 (14th)	118 (18th)	371 (21st)	267 (29th)				134 (22nd)	90 (20th)
Vandergrift Bridge		498 (14th)*	340 (24th)	317 (23rd)	175 (5th)	108 (4th)	73 (25th)	55 (14th)	127 (18th)	368 (21st)	284 (29th)	238 (16th)	149 (28th)	90 (11th)	128 (22nd)	83 (20th)
POTW C							1.2 X							1.1 X		
Leechburg Bridge		489 (14th)**		177 (23rd)	143 (5th)	108 (4th)	84 (25th)	57 (14th)	126 (18th)	316 (21st)	271 (29th)	191 (16th)		102 (11th)	106 (22nd)	83 (20th)
Kiski Railroad Bridge	850 (24th)***	500 (14th)***	109 (30th)***	181 (23rd)	219 (5th)				113 (18th)	300 (21st)	274 (29th)	186 (16th)	146 (28th)	116 (11th)	122 (22nd)	

Notes:

*Vandergrift (RDB)

**Leechburg(RDB)

***Kiski Camp Site(RDB)

RDB = Right Descending Bank

LDB = Left Descending Bank

Table 9
Lower Allegheny River
Bromide Concentration (ug/l)

Sample Site	Sept 2010	Oct 2010	Oct 2010	Nov 2010	Dec 2010	Jan 2011	Feb 2011	March 2011	April 2011	May 2011	June 2011	July 2011	Aug 2011	Sept 2011	Oct 2011	Nov 2011	Dec 2011
Water Plant Intake(RDB)	170 (24th)*	115 (14th)*		72 (30th)	76 (28th)		137 (4th)	30 (25th)	40 (14th)	26 (18th)	90 (21st)	220 (28th)	202 (16th)	124 (28th)	107 (11th)	66 (22nd)	64 (20th)
River Forest Yacht Club (LDB)		155 (14th)		96 (30th)	134 (28th)			60 (25th)	47 (14th)	44 (18th)	123 (21st)	203 (28th)	155 (16th)	144 (28th)	110 (11th)	98 (22nd)	62 (20th)
Buffalo Creek (AMD)																	
Veterans Road (RDB)											113 (17th)	150 (18th)	142 (17th)	86 (23rd)	72 (7th)	96 (17th)	66 (19th)
Steel Plant A												1.1 X	1.6 X	1.2 X			
Tarentum (RDB)	220 (24th)	158 (15th)		158 (24th)			62 (25th)	34 (16th)	34 (19th)	43 (27th)	112 (17th)	161 (18th)	231 (17th)	104 23rd	69 (7th)	70 (17th)	64 (19th)
Coal Fired Power Plants C & D				1.2 X					1.2 X			1.2 X		1.2 X			1.3 X
Rachel Carson Park (RDB)							48 (25th)	34 (16th)	40 (19th)	49 (27th)	116 (22nd)	196 (18th)	170 (17th)	79 (23rd)	69 (7th)	77 (17th)	85 (19th)
Harmar Marina (RDB)	230 (24th)	149 (15th)		190 (24th)	65 (23rd)	122 (27th)		35 (8th)	55 (7th)	43 (17th)	113 (21st)	183 (28th)	176 (18th)	132 (28th)	111 (13th)	67 (22nd)	73 (20th)
Harmar Mine (AMD)																	
POTW D																	
Hulton Bridge (CTR)	220 (24th)	139 (15th)	205 (29th)	191 (24th)	51 (23rd)	128 (27th)		<25 (8th)	49 (7th)	40 (17th)	101 (21st)	184 (28th)	168 (18th)	126 (28th)	102 (13th)	61 (22nd)	70 (20th)
Hulton Bridge (RDB)	210 (24th)		221 (29th)	202 (24th)	63 (23rd)	133 (27th)		<25 (8th)	56 (7th)	42 (17th)	111 (21st)	191 (28th)	164 (18th)	141 (28th)	99 (13th)	64 (22nd)	70 (20th)
PWSA Intake (RDB)	220 (24th)	151 (15th)	241 (29th)	204 (24th)	79 (24th)	130 (27th)		<25 (8th)	49 (7th)	38 (17th)	94 (21st)	177 (28th)	172 (18th)	107 (28th)	122 (13th)	63 (22nd)	59 (20th)
Lock & Dam #2 (LDB)	230 (24th)	147 (15th)	263 (29th)	213 (24th)	62 (23rd)	141 (27th)		<25 (8th)	52 (7th)	45 (17th)	126 (21st)	180 (28th)	161 (18th)	117 (28th)	109 (13th)	56 (22nd)	64 (20th)

Notes:

*Freeport(RDB)

RDB = Right Descending Bank

LDB = Left Descending Bank

CTR = Center

AMD = Acid Mine Drainage

Table 10

Bromide Mass (pounds/day) Input to the Allegheny River System (*sum of upstream contributions from Industrial Waste Treatment Plants A,C & D and Coal Fired Power Plants A & B.)

Sampling Location	January	February	March	April	May	June	July	August	September	October	November	December
Franklin Bridge	2,377	5,018	2,870	5,989	6,927	1,760	1,021	1,454	2,437	917	4,690	2,212
Industrial Waste Plant A Mass Added												
Kennerdell Bridge	4,716	13,337	5,855	10,731	6,597	2,903	2,737	1,640	1,533	3,314	5,554	793
Route 56 Bridge-Armagh	56	75	74	95	89	42	30	46	201	60	90	134
Industrial Waste Plant C Mass Added												
Route 119 Bridge(Jan-Jun); Newport Road Bridge (July-Dec)	622	606	303	460	498	888	621	78	810	58	1,162	169
Route 56 Bridge-Johnstown	180	362	383	417	200	72	14	157	152	194	218	250
Coal Fired Power Plants A&B Mass Added												
Seward Bridge	190	362	383	464	1,007	520	61	936	1,401	194	218	607
Conemaugh Dam Trail		2,974	5,148	1,603	2,739	803	2,348	2,138	845	2,777	2,025	2,574
Industrial Waste Plant D Mass Added												
Tunnelton Road Bridge		2,793	6,436	1,984	4,108	956	2,621	2,764	972	3,057	2,145	1,802
Sum of the Measured Mass Added to the Allegheny River System*	2,915	8,669	4,501	5,535	2,256	2,591	2,626	1,622	1,081	2,676	2,056	-1,798
Bromide Mass at PWSA Intake — Minimum	3,687	6,640	5,902	5,537	4,317	2,979	2,245	3,309	2,154	3,396	4,608	4,088
Bromide Mass at PWSA Intake — Maximum	28,507	34,313	35,013	28,510	18,012	8,049	8,090	6,322	16,409	16,256	19,897	19,731
Bromide Mass at PWSA Intake — Mean	8,884	11,708	14,209	14,049	9,970	4,299	3,623	4,632	7,164	9,065	9,705	8,143
Mean bromide mass at the PWSA intake minus bromide mass at Franklin Bridge	6,507	6,690	11,339	8,060	3,043	2,539	2,601	3,178	4,728	8,148	5,015	5,931
Percentage of observed bromide mass change measured from industrial sites compared to the bromide mass difference between the Franklin Bridge and PWSA intake	45	130	40	69	74	102	101	51	23	33	41	-30

MANAGEMENT OF SOLUBLE ORGANICS IN PRODUCED AND FLOWBACK WATERS WITH SWELLABLE, ABSORBENT GLASS

Paul Edmiston[1], Justin Keener[2*], Shawn McKee[2], Scott Buckwald[2], Gregory Hallahan[2], Michael Grossman[2]

[1]The College of Wooster, ABSMaterials, Inc., Wooster, Ohio
[2]Produced Water Absorbents, Inc., Wooster, Ohio
*Email: j.keener@pwabsorbents.com

ABSTRACT

Osorb®, an absorbent, nano-engineered glass, is being commercialized to remove soluble organics like BTEX and other volatile organic compounds (VOCs) from oilfield produced and flowback waters for the purposes of beneficial reuse, membrane protection, air emission controls, and offshore excursions. The Osorb is reusable, utilizing one or more of centrifugal, thermal, and vacuum regeneration processes. Two pilot water treatment systems are currently being fielded, and a commercial 5 bpm system is undergoing final engineering. PWUnit#1.5 is a fully automated, mobile system that polishes soluble organics from water at 246 L/min. The design parameters for PWUnit#1.5 are based on the manually operated PWUnit#1. Final engineering is being completed for the 798 L/min PWUnit#2, which includes closed-loop water treatment and Osorb regeneration. Excursion Unit#1 is a 160 L/min, fixed-bed system designed for polishing soluble organics from produced water prior to overboard discharge during offshore excursions.

KEYWORDS: Osorb®, Swellable, Absorbent, Glass, Produced Water, Flowback, Excursion, Soluble Organics, BTEX, VOC, Hydrocarbons, PWUnit

INTRODUCTION

With the increasing utilization of hydraulic fracturing and the generation of trillions of liters of produced water each year, management of produced water to maintain environmental protection and economic efficiency is one of the oil and gas industry's greatest challenges. This will be especially vital as the U.S. continues to rapidly develop its domestic shale formations. From a regulatory standpoint, both water and air regulations dictate water management strategies to maintain to water quality requirements and air pollution standards. Soluble organics must be removed prior to surface discharge or agricultural application, and volatile organic compounds (VOCs) can lead to smog formation if not removed prior to storing water in open tanks or evaporation ponds.

Beyond regulations, the application of filtration or desalination processes such as reverse osmosis often require protection of costly membranes by removing soluble organics to prevent bacteria growth.

The highly variable contaminant profile of this water requires technologies that can treat a wide range of species while withstanding the harsh chemical environment. Produced Water Absorbents, Inc. is using Osorb®, an absorbent glass, in a suite of continuous and excursion treatment systems as a polishing step to remove soluble organics from produced waters and flowback waters. While many technologies are available to remove freely dispersed hydrocarbons, very few have the ability to effectively capture the soluble organics from these contaminated water streams. To fill this void, two pilot treatment systems and a pilot regeneration system are currently being fielded, and the first commercial-scale system has entered the final stages of engineering. This fleet of systems consists of:

- PWUnit#1.5 - An automated, weatherized pilot system that is integrated into a 53' trailer and treats water at 246 L/min utilizing a second-generation treatment process based on the earlier pilot system, PWUnit#1.
- RegenUnit#1 is a mobile Osorb regeneration system that was originally deployed with PWUnit#1 and utilizes a combination of agitation, heat, and vacuum to remove the captured organics from the Osorb prior to reusing the Osorb in PWUnit#1.5
- PWUnit#2 is being designed to treat water at 798 L/min, utilizing a closed-loop water treatment and Osorb regeneration process consisting of dissolved air flotation, centrifugation, and thermal polishing.
- Excursion Unit#1 - A 160 L/min system on a 2 m^2 skid that uses interchangeable Osorb cartridges to polish soluble organics from produced water on offshore platforms.

TECHNOLOGY OVERVIEW

Osorb® is a hybrid organic-inorganic material based on silica that can swell to up to eight times in mass in the presence of neat organics.[3-6] Osorb is hydrophobic and does not absorb water, but is effective at removing a wide range of soluble organics from water. Sorbate compounds range from polar species such as alcohols, carboxylic acids, and non-ionic surfactants to non-polar species such as toluene, benzene, ethylbenzene and decane.[7] One of the most unusual aspects of Osorb, which on the macroscopic level is a glass-like material, is the ability to rapidly swell. Swelling is reversible and occurs in <1 s when exposed to neat organic liquids (Figure 1). There is no loss in the observed swelling behavior even after repeated use. Remarkably, the swelling resulting from absorption of neat liquids is so energetic that the material expands with forces in excess of 500 N/g which allows the material to lift objects over 40,000 times heavier in mass. Chemically, swellable glass is composed of polymerized bis(trimethoxy-silylethyl)benzene, making it similar in composition to poly(dimethylsiloxane) but possessing a bridging aromatic group which has been shown to be a key component in generating the swelling behavior.[4]

The absorption process of Osorb has been extensively studied in previous research and detailed elsewhere.[3] Briefly, capture of organic compounds is due to mechanical expansion of a collapsed matrix of silica nanostructures arranged as a complex and microscopically disorganized nanoporous network (Figure 1). Capillary-induced collapse of the nanoporous matrix is incurred upon evaporation of solutes. The tensile forces created in the contraction of the matrix are stored as non-covalent interactions within a high internal surface area (>800 m^2/g), preventing re-expansion. Upon absorption, the swellable glass relaxes to the expanded state creating new surface area and volume for subsequent molecules to be absorbed and permeate through the nanopores (Figure 2). Water is too polar to enter the hydrophobic pore structure, and absorption affinity of individual chemical compounds trends with polarity as measured by the octanol-water partition coefficient, $\log K_{ow}$. In general, the partition coefficient for a dissolved organic species between water and swellable glass is approximately one order of magnitude greater than $\log K_{ow}$. Larger partition coefficients are attributed to relaxation via expansion of the internal pore structure. Pore expansion also leads to high capacity and pseudo-linear absorption isotherms through all ranges of concentration.[3]

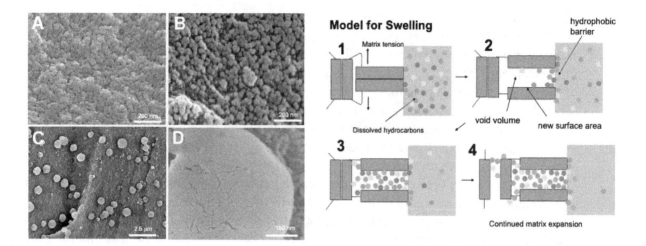

Figure 1. Electron micrographs of: A. Unswollen Osorb (1 polymerized in THF using TBAF catalysis). B. Osorb swollen in a solution of poly (2,2,3,3, 4,4,4-heptafluorobutyl-methacrylate) and dried to leave entrapped polymer. C & D. Osorb swollen in ethanol followed by critical point drying showing the fully expanded state.

Figure 2. Proposed model for absorption of dissolved organics by Osorb media. 1. Initial adsorption to the surface of the material. 2. Sufficient adsorption occurs to trigger matrix expansion leading absorption across the sorbent-water boundary. 3. Pore filling leading to further percolation into the nanoporous matrix. 4. Continued matrix expansion increases available void volume.

During the past 12 months new formulations of Osorb have been developed, some which have been proven effective at treating highly contaminated produced water. The base formulation of Osorb has a particularly high affinity for BTEX, gasoline-range organics (GRO), and diesel-range organics (DRO). While the highest affinity is observed for these non-polar organics, Osorb has exhibited the unique ability to also extract polar soluble compounds such as methanol. While the high affinity for BTEX and other hydrocarbons is generally independent of any environmental factors, the ability of Osorb to extract methanol is affected by the concentration of dissolved solids and presence of co-sorbates in the water. The increased concentration of ions in high salinity produced water disrupts hydrogen bonds between methanol and water molecules, allowing methanol to be extracted from the water by Osorb with greater efficiency. Addition of polar or ionic polymers into the Osorb matrix can improve the extraction of protic organics such as methanol and phenol by facilitating transport across the water-hydrophobic pore boundary.

BENCH TESTING AND DATA

Analytical Methods

Analytical methods on treated and untreated samples of produced water were analyzed using a modified ISO 9377-2 method with dichloromethane extraction for quantitation of BTEX, hydrocarbons, and TPH. Briefly, 20 mL of each sample is extracted with 1.0 mL of dichloromethane and the extract is analyzed by a gas chromatography (GC) – flame ionization detector (FID) using a Rtx-5 column. Mineral oil is used as a standard for oil & grease while individual standards of BTEX compounds are used. Direct aqueous injection (GC) was used for methanol analysis using a Stablewax column with FID detection.

Bench Extraction Data

A variety of tests using Osorb on single component and multi-component water samples have been conducted on the bench-scale to understand the capabilities of Osorb towards the extraction of dissolved species found in produced water (Table 1). Additional testing has been conducted on various produced water samples to analyze the effects of varying the amount of Osorb (%w/v) used for treatment. These tests have indicated that 0.2-1.0% w/v Osorb is effective at capturing dissolved BTEX and hydrocarbons from produced water streams. A standard Osorb amount of 0.5% w/v of water has been used as a benchmark for water treatment to conduct most bench-scale testing, including the data obtained in Table 1.

Table 1. Effectiveness of Osorb for extraction of various organic compounds.

Compound	% Extraction
Aliphatics	
Straight chain	>99%
Branched chain	>99%
Cyclic	>99%
Aromatics	
BTEX	>98%
Polyaromatic	>99%[#]
Acids	
Carboxylic acids	40-95%, Depends on chain length
Napthenic acids	>90%
Aromatic acids	10-90% (depends on pH)
Phenol	>50% (pH =7.0) >90% pH
Surfactants	
Polyethylene oxide	>90%
Triton X-100	>90%
Alkyl glucosides	50-60%%
Other	
Metals	>80% extraction Fe^{2+}, Hg^{2+}, Ba^{2+} and Pb^{2+}
Polymers	up to 90% extraction depending in class (inquire)
Alcohols	4-90%, depends highly on salt concentration

Bench Extraction of Methanol and Phenol

Extraction of dissolved alcohols from water is a significant challenge in the treatment of produced water. Osorb has the unique ability to capture polar solubles like methanol from water which is hypothesized to be due to the thermodynamics of matrix relaxation aiding absorption. Extraction efficiency is dependent on salt concentration, thus all testing was done at a standard condition using 2,000 ppm methanol; 50 ppm BTEX, 7,000 ppm NaCl, and 100 ppm $CaCl_2$ (7,100 TDS). Three replicates were performed for all tests. Methanol was measured by direct injection gas chromatography with flame ionization detection. Internal standards were used. Although not reported below, extraction of BTEX was >90% in all instances. In order to alleviate the dependency on TDS concentrations for removal of polar organics, new formulations of Osorb with modified surface chemistries have been developed. Table 2 provides data for the extraction of methanol and phenol using basic Osorb and modified Osorb formulations.

Table 2. Extraction of methanol and phenol by Osorb materials

Sorbent Type	Percent Reduction[#]		
	Methanol (single step)	Methanol (multi-step)*	Phenol
Osorb	8.2±3.8	25.0±3.8	10±3.8
Iron-Osorb	16.1±1.3	38.2±1.2	43±2.0
PolyAnion-Osorb	11.2±1.3	40.8±3.6	<1
PolyCation-Osorb	4.8±1.3	n/m	58±0.5

* Four successive extractions with 0.5% w/v Osorb
[#] Extraction using 0.5% w/v Osorb, 3 min contact time, T = 25°C
Methanol = 2,000 ppm, phenol = 100 ppm

An interesting result from this work was the observation that methanol extraction is increased by as much as 500% when co-absorption with BTEX occurs. Understanding the origin of this enhancement is currently being explored.

Bench Extractions of Produced Water

A sample of Wyoming produced water was collected for bench scale testing. Oil sheen was visible on top of the water sample, indicating the presence of high hydrocarbon concentrations. The sheen was removed prior to treatment. Osorb was highly effective at removing the organic components from the produced water. Using 2% w/v Osorb (kg media/L water) with 60 seconds of contact time resulted in ~98% extraction of organic species as measured by gas chromatography-mass spectrometry. As suggested by the visible sheen, the majority of the organic components were hydrocarbons. The post-treatment results (Figure 3) show only two detectable compounds remaining after extraction. These were two polar surfactants (ex. peak at 5.8 min is 2-butoxy-ethanol) that were ~85% extracted. Depending on the concentrations and species of organics, especially dispersed organics, larger amounts of Osorb may be required. For the sample of Wyoming produced water described in Figure 3, application of 0.5% w/v Osorb maintained a greater than 98% contaminant removal.

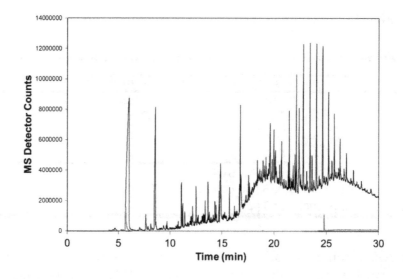

Figure 3. Gas chromatography-mass spectrometry analyses of (black) untreated produced water and (red) water after 60 second treatment with 2% w/v Osorb.

Dependence on Temperature

The extraction of 125 ppm dissolved toluene from water was measured at elevated temperatures. Testing was accomplished using laboratory samples at TDS = 0 ppm and TDS = 1,000 ppm with 0.5% w/v Osorb (Table 3 and Table 4). There is a decrease in extraction when T = 85°C; however, percent extraction exceeds 90%.

Table 3. Extraction of dissolved toluene (125 ppm) from deionized water

Temperature (°C)	Percent Extraction*
20	97.20%
70	95.10%
85	91.30%

*error +/-2%

Table 4. Extraction of dissolved toluene (125 ppm) from salt water (TDS = 1,000 ppm NaCl)

Temperature (°C)	Percent Extraction*
70	99.50%

*error +/-2%

To test the effectiveness of Osorb at decreased temperatures, a sample of this Wyoming produced water was chilled to slightly above freezing (33.5 F). The standard 2 g of Osorb was used to treat 20 mL of produced water with 30 seconds of contact time. The ISO 9377 method detailed earlier was used for extraction and analysis. The resulting spectra indicated an extraction efficiency of 99.18%, suggesting that cold temperatures do not inhibit the absorption of contaminants.

Bench Scale Removal of High Molecular Weight Polymers from Flowback

Substantial work has been completed on the development of new materials to absorb the larger molecular weigh organics that are characteristic to flowback water. The base formulation of Osorb has a small surface pore structures when fully collapsed (~10 nm) which sterically inhibit the transport of polymers into matrix, especially when the concentration of other organic absorbates is low (co-absorption helps to open the pore structure). New versions of Osorb with altered chemical structure and morphology were designed to open the pore structure and allow for capture of polymer species such as polyacrylamides. The best material developed in this effort is termed Osorb-EB (EB = emulsion breaker) and is created using silanes that are bridged by alkyl groups. Using nitrogen adsorption porimetry, the surface area of Osorb-EB is 980 m^2/g with an internal pore volume of 1.14 mL/g, both of which are approximately twice that of base Osorb. Initial testing of Osorb-EB moved directly to treatment of actual flowback water and other hard to treat waters in oil/gas production. Osorb-EB is was found to be particular well suited to treat flowback water when introduced into the fluid with a high speed agitated mix to facilitate capture. The material floats and has the ability to adhere to particles resulting in the removal of organics and solid particles (Figure 4). TOC indicates that the organic carbon levels were reduced from 265 ppm to <5 ppm.

127

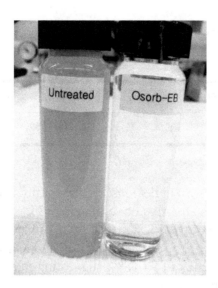

Figure 4. Flowback water before and after treatment with 0.4% w/v Osorb-EB.

PILOT TREATMENT SYSTEMS

PWUnit#1

The trailer-mounted PWUnit#1 was designed to remove soluble organics from produced water at 246 L/min. As produced water enters the system, it first passes through a single filtration vessel to remove any residual solids. A mixing eductor and volumetric feeder are then used to continuously add Osorb into the water stream. The produced water and Osorb are passed through two agitation tanks in series, where a combination of mechanical agitation and eductor nozzles are applied. The combination of the agitator and mixing eductors create sufficient interaction between the Osorb and the solubles in the water for absorption. Following the capture of solubles inside the mixing vessels, the Osorb and treated water pass over a vibratory separator to collect the organic-laden Osorb from the water discharge stream. As the water is discharged to the target location, a filtration manifold removes any fine Osorb particles that may have passed through the vibratory separator screen.

During pilot testing in Spring 2011, PWUnit#1 reduced an average TPH concentration of 245 mg/L in Clinton formation produced water to <0.1 mg/L (Table 5). This is consistent with data collected during testing in Summer 2010, when the system reduced TPH concentrations in Clinton Formation produced water from 277 mg/L to <0.1 mg/L.

Table 5. Data collected during the treatment of Clinton Formation produced water with PWUnit#1 in Spring 2011.

Sample #	TPH Concentration (ppm)		TPH Removal Efficiency (%)
	Untreated	Treated	
1 (U) & 3 (T)	261	< 0.1	99
4 (U) & 5 (T)	218	< 0.1	99
7 (U) & 9 (T)	255	< 0.1	99

(U) Untreated & (T) Treated

RegenUnit#1

RegenUnit#1 consists of a double cone vacuum dryer, thermal fluid controller, vacuum pump, and condenser. After capturing solubles from produced water in one of the pilot treatment systems, Osorb is transferred into the double cone vacuum dryer. The dryer is agitated, heated, and evacuated in order to drive the organics from the Osorb matrix. The dryer vessel is rotated at 6 rpm to agitate the Osorb and is heated by the thermal fluid controller, which circulates fluid through an outer jacket. The vacuum pump pulls the organic vapors from the dryer and passes them through the condenser, where the organics are condensed and collected. The collected organics are either properly disposed of or returned to the client for further processing or resale. If high boiling point solubles have been collected by the Osorb, solvent can be used during the regeneration process to rinse the Osorb. After regeneration in RegenUnit#1, the Osorb is available for reuse in the pilot treatment systems to treat additional produced water. PWUnit#1 and RegenUnit#1 were deployed for pilot testing in Wamsutter, Wyoming, achieving greater than 93% BTEX reduction in produced water from natural gas wells (Figure 5).

129

Figure 5. PWUnit#1 and RegenUnit#1 set up for pilot treatment of produced water in the Wamsutter gas fields near Wamsutter, Wyoming.

PWUnit#1.5

PWUnit#1.5 (Figure 6) is an automated, weatherized system that treats water at 246 L/min utilizing a treatment process founded on the design of PWUnit#1. As produced water enters the system, solids are filtered using an automatic backwashing sand filter. Following prefiltration, PWUnit#1.5 uses a mixing eductor, volumetric feeder, and vacuum conveyor to continuously add Osorb into the produced water stream. Similar to PWUnit#1, the water is processed in a series of two mixing tanks; however, both mixing tanks utilize dual-impeller agitators to mix the Osorb with the water, allowing the Osorb to efficiently capture the solubles while maintaining the integrity of the Osorb media. Following the water treatment in the mixing tanks, the water and Osorb pass over a vibratory separator to separate the water and organic-laden Osorb. The Osorb is transferred to the RegenUnit#1 for regeneration, and the treated water is passed through a manifold of filter vessels to capture any fine particles of Osorb that may have passed through the vibratory separator screen.

PWUnit#1.5 is integrated into a 53' trailer for mobility and on-board laboratory analysis. A retractable vinyl covering is used to weatherize the system for operation in wet or cold climates. Wet testing is underway in Third-Quarter 2012, with fielded pilots expected by the end of the quarter. Figure 7 depicts the process flow diagram of PWUnit#1.5 and RegenUnit#1 when configured for continuous processing.

Figure 6. PWUnit#1.5 prior to initial wet testing. PWUnit#1.5 is designed to continuously polish soluble organics from produced water at a rate of 246 L/min.

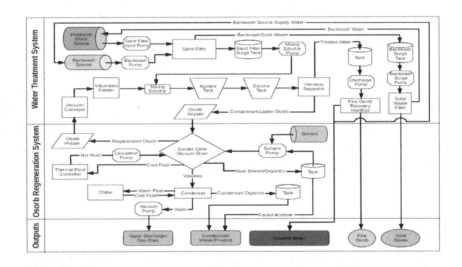

Figure 7. The process flow diagram of PWUnit#1.5 and RegenUnit#1 when configured to continuously polish solubles from produced water and regenerate Osorb.

COMMERCIAL SYSTEM DEVELOPMENT - PWUNIT#2

PWUnit#2 is being engineered as the first commercial Osorb-based system for polishing of soluble organics and will consist of a closed-loop cycle of 798 L/min water treatment and Osorb regeneration. The system will include a complete automation package with in-line analytical equipment and remote monitoring for off-site operation. After the completion of fielded pilot testing in 2011, Fourth-Quarter 2011 and First-Quarter 2012 were dedicated to extensive research and development for PWUnit#2. This testing regime included analyses of regeneration methods, Osorb application methods, fixed media vessel design, and Osorb's physical properties. During the pilot testing in Wyoming, it was observed that regeneration of Osorb required 8-12 hours in RegenUnit#1 due to extensive retention of water on the surface of Osorb after use. Due to the time requirement for regeneration and the rate of Osorb usage in PWUnit#1, the treatment of water was frequently paused to allow for regeneration of the backlogged Osorb. In order for PWUnit#2 to incorporate a closed-loop cycle of water treatment and Osorb regeneration, a regeneration method was required that maintained a consistent pace with the generation of organic-laden Osorb from water treatment. It was found that application of centrifugation removed greater than 90% of the residual surface water on Osorb. Additionally, an average of 50-70% of absorbed organics was removed with the surface water. As a result of this testing, it was concluded that a combination of centrifugation and a polishing dryer would drastically decrease regeneration time and energy requirements and accommodate the closed-loop design of PWUnit#2.

During this period of testing, a manifold was designed that can provide sufficient mixing of the Osorb and water while alleviating the footprint and scale-up issues associated with the mixing tanks on PWUnit#1 and PWUnit#1.5. On the output of this mixing manifold, dissolved air flotation (DAF) proved to be the best method to separate the Osorb and water due to the density and hydrophobic character of Osorb. Not only are DAF units more scalable than the vibratory separators used on PWUnit#1 and PWUnit#1.5, but DAF can effectively capture solid particles down to approximately 20 micron in size. As the particle size of Osorb decreases, the absorption effectiveness and ease of regeneration increase. PWUnit#2 will therefore not only accommodate higher flow rates with shorter regeneration time requirements, but the extent of solubles removal from the contaminated water streams will also improve in comparison to the results achieved with PWUnit#1.

EXCURSION SYSTEMS

Currently, regulations for offshore oil and gas operators in the Gulf of Mexico restrict the level of organics that can be discharged overboard during an increased influx of produced water, drilling mud, or other water source needed during exploration and development. These levels are set at 29 mg/L maximum average for monthly discharge, and 42 mg/L maximum average daily discharge.[8] Smaller platforms discharge up to 160 m^3 water per day, while larger platforms can discharge upwards of 6500 m^3 water per day. Discharges with elevated concentrations of organics are subject to federal fines or mandates to shut in wells. They also cause public scrutiny of drilling and production practices offshore in environmentally sensitive habitats. Various technologies already exist offshore to eliminate much of the organics from these overboard discharges, but modern technology currently lacks the ability to consistently decrease concentrations to less than 1 mg/L and remove the water soluble organics (WSO).

A 160 L/min pilot system, Excursion Unit#1 (Figure 8), has been constructed to utilize Osorb cartridges as a polisher for soluble organics before final discharge of produced water into the GOM. The system consists of four parallel Osorb cartridges and is maintained within a 2 m^2 footprint. Produced water tagged for overboard discharge passes through a solid/liquid separator and an oil/water separator, which typically results in dispersed and dissolved organic concentrations below the regulated limits. In the event where the primary treatment system fails to meet discharge standards, fluid would pass through Excursion Unit#1 to polish the elevated organics to meeting regulations. The cartridge design allows for rapid exchange once the material has reached capacity and experiences breakthrough. The Osorb cartridges are then sent back to onshore processing locations for regeneration of the Osorb prior to reuse.

Figure 8. Excursion Unit#1 during wet testing in Wooster, Ohio in June 2012

Wet testing of Excursion Unit#1 has confirmed that the system can effectively capture 99% of dissolved toluene at 100 mg/L for up to 10 hours before reaching breakthrough. Breakthrough is defined as less than 99% capture. Through Third-Quarter 2012, Excursion Unit#1 will be wet testing with a broad range of soluble and insoluble organics to determine breakthrough curves and loading capacities. A concurrent series of bench-scale testing will be used to validate the results obtained during wet testing.

CONCLUSIONS

The combination of increased development of shale plays and increased overall global oil and gas production is strengthening the requirement for technologies that can manage the produced and flowback water streams from this development. A high-capacity, reusable media has been developed to effectively polish soluble organics from these waters. Numerous bench tests have validated the effectiveness of the technology, including under near-freezing conditions and TDS levels over eight times more concentrated than seawater. Life-cycle testing indicates that the effectiveness of the material does not diminish after repeated capture of solubles. Additionally, modified Osorb formulations have been developed for the purpose of removing highly soluble polar organics from produced water.

In 2011, the 246 L/min PWUnit#1 and RegenUnit#1 completed multiple pilot tests on Clinton formation produced water, consistently reducing TPH levels to less than 0.1 mg/L. Deployment of these systems for pilot testing in Wamsutter, Wyoming to treat produced water from natural gas wells resulted in consistent reduction of BTEX levels by over 93%. These pilot efforts led to the fabrication of PWUnit#1.5, which streamlines the PWUnit#1 246 L/min treatment process and upgrades the system to include integration into a single trailer, automation, and weatherization. PWUnit#1.5 is undergoing wet testing in Third-Quarter 2012.

Engineering of PWUnit#2 is nearing completion, using lessons learned from the 2011 pilot program, construction of PWUnit#1.5, and the series of tests run during Fourth-Quarter 2011 and First-Quarter 2012. The system will pass contaminated water through a mixing eductor and mixing manifold to absorb soluble organics with Osorb, and the Osorb will then be recollected using dissolved air flotation. Continuous regeneration of the Osorb will be achieved using a centrifuge and polishing dryer, creating a closed-loop cycle of Osorb between water treatment and regeneration.

In addition to these onshore applications for produced water and flowback water treatment, Excursion Unit#1 has been developed to provide operators with a polishing unit for their overboard discharge waters. Wet testing of Excursion Unit#1 began June 2012 and will be continued through Third-Quarter 2012.

ACKNOWLEDGMENTS

The authors gratefully acknowledge the assistance of the ABSMaterials and Produced Water Absorbents laboratory and operations staff for their assistance in sample analysis, technology development, and successful deployment of pilot systems. Additionally, the authors would like to recognize the following organizations for their support: National Science Foundation, United States Department of Energy, Ohio Environmental Protection Agency, United States Environmental Protection Agency, Army Corps, Texas A&M University, Global Petroleum Research Institute, Energy Ventures, Harris & Harris Group, and The College of Wooster.

REFERENCES

1. Veil., J. A.; Puder, M. G.; Elcock, D.; Redweik, Jr., R. J. (2004) A White Paper Describing Produced Water from Production of Crude Oil, Natural Gas, and Coal Bed Methane.

2. Colorado School of Mines. (2009). An Integrated Framework for Treatment and Management of Produced Water. *Technical Assessment of Produced Water Treatment Technologies, 1st Ed.* RPSEA Project 07122-12.

3. Edmiston, P. L.; Keener, J.; Buckwald, S.; Sloan, B.; Terneus, J. (2011). Flow Back Water Treatment Using Swellable Organosilica Media. SPE-148973-PP.

4. Burkett, C. M., Edmiston P. L., Highly swellable sol-gels prepared by chemical modification of silanol groups prior to drying. J Non-Crystalline Solids 2005 (351):3174-3178.

5. Edmiston, P. L. US Patent 1/537,944 US 2007/0112242 A1 Swellable Sol-Gels, Methods Of Making, And Use Thereof.

6. Edmiston, P. L., Underwood, L.A. Absorption of dissolved organic species from water using organically modified silica that swells. Separation. Purification. Technol. 2009 (66) 532-540.

7. Burkett, C. M. Underwood, L. A., Volzer, R. S., Baughman, J. A., Edmiston, P. L. Organic-inorganic hybrid materials that rapidly swell in non-polar liquids: Nanoscale morphology and swelling mechanism. Chem Mater. 2008 (20): 1312-1321.

8. National Energy and Technology Laboratories. Produced Water Management Information System Federal Regulations: U.S. Environmental Protection Agency. http://www.netl.doe.gov/ technologies/pwmis/regs/federal/epa/index.html

WASTEWATER TREATMENT CHALLENGES ASSOCIATED WITH NONCONVENTIONAL OIL AND GAS ACTIVITY IN PENNSYLVANIA

Daniel Ertel[1], Kent McManus[2], Timothy Butters[1], Jerel Bogdan[2]

[1] Eureka Resources, LLC, 419 Second Street, Williamsport, PA 17701, USA

[2] ARCADIS US, 50 Fountain Plaza, Suite 600, Buffalo, NY 14202, USA

ABSTRACT

Thanks to the development of new nonconventional technology, the development of huge pools of oil and gas reserves, once thought to be too deep and held too tightly in rock, is now happening at a feverish pace in several areas or shale basins of the world. One of the most prominent of these basins is the Marcellus Shale Basin – underlying large swaths of New York, Pennsylvania, Ohio, West Virginia, and Maryland. Of the many challenges associated with shale gas development, management of water used and wastewater generated is one of the most prominent. This presentation will focus on the technical challenges the industry is facing regarding management and treatment of wastewater generated from both drilling and completions (including "hydrofracturing") operations associated with development in the Marcellus Basin, including challenges associated with hard-to-remove pollutants, unit processes required, disposal and reuse options and limitations, and regulatory drivers and restrictions.

A case study will be presented focused upon Eureka Resources, LLC privately-owned and operated wastewater treatment facility in North Central Pennsylvania that is employing advanced treatment of these wastewaters to meet the needs of developers and drillers, all the while balancing compliance with strict environmental regulations in Pennsylvania. The case study will focus on technical challenges faced by management and operations staff as they deal with providing effective treatment of wastewater with ever-changing influent characteristics and unique constituents. Specifically, the case study will include discussion of:

The characteristics of influent to their facility and how influent variability affects their treatment strategy and operations.

- The overall treatment strategy and unit processes employed by Eureka at their facility, including mechanical vapor recompression distillation treatment systems.
- Challenges pertaining to management of concentrated brine produced by their facility.
- How Eureka is staying ahead of the regulatory curve.

137

KEYWORDS: Shale Gas Wastewater, Development Water, Produced Water, Treatment, Marcellus, Reuse, Recycle, Case Study

INTRODUCTION

Natural gas represents an important energy source for the United States (US). According to the US Department of Energy's (DOE's) Energy Information Administration (EIA), over 20% of the country's energy needs are provided by natural gas. Thanks to the development of new nonconventional technology, the development of huge pools of oil and gas reserves, once thought to be too deep and held too tightly in shale rock formations, is now happening at a feverish pace in several areas of the US and the world. One of the most prominent of these "basins" in the US is the Marcellus Shale Basin – underlying large swaths of New York, Pennsylvania, Ohio, West Virginia, and Maryland. In the highly competitive US shale gas market, it is important for oil and gas (O&G) exploration and production (E&P) companies to find efficient solutions to one of the high-cost aspects of production; treatment of flowback and produced waters. Due to scarcity of water resources as well as regulatory constraints and expensive treatment technologies, E&P companies must develop and implement innovative methods for re-use and adopt more localized and centralized water management systems. E&P companies in various shale plays across the US are now consistently seeking out new methods to **transport, dispose,** and **re-use** flowback and produced water to maximize the profitability of shale gas production and remain competitive in a fast-growing market.

This paper covers the following topics with a focus on the Marcellus Shale Gas play in Pennsylvania:

- Flowback/Produced Water Management/Treatment Challenges
- Typical Characteristics of Marcellus Shale Gas Play Wastewaters
- Regulatory Drivers
- Eureka Resources Case Study
- Conclusions/Takeaways

Flowback/Produced Water Management/Treatment Challenges

E&P companies face many challenges when developing and implementing water management/treatment/ disposal strategies including variable water quality, need to establish treatment objectives, need to evaluate treatment technologies/schemes and need to adjust the approach to address operational issues and regulatory requirements (see Figure 1). The types of O&G wastewater streams in the Marcellus play, along with a summary of primary constituents of interest that may require management/treatment are summarized in Table 1.

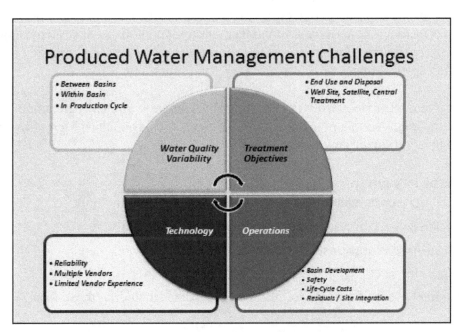

Figure 1 – Produced Water Management Challenges

Table 1

Oil and Gas Wastewater Types and Primary Constituents of Interest

Top Hole	Relatively Clean
Pit Water - Drilling Fluids/Muds/Formation Water	Total Suspended Solids (TSS), Chemical Additives
Flowback (Chemical Additives,	Total Dissolved Solids (TDS), Metals, Naturally Occurring Radioactive Materials (NORM)
Produced	TDS, Metals, NORM
Condensate	Oil, Benzene/Toluene/Ethyl Benzene and Xylene (BTEX)
Drilling Muds	TDS, NORM, TSS, Chemical Additives, Oil (if oil based mud)

Flowback and produced waters are typically the larger volume wastewater streams. In general, the relative volumes of these two wastewater streams typically change over the life of a shale gas play development as shown on Figure 2, with flowback flows decreasing and produced water flows increasing as production activity outpaces drilling and completions activities. The structure and rate of this change is hard to predict and difficult to plan for; however, it is one of the factors that must be considered when developing an effective long-term water management/treatment/disposal strategy. This paper uses the term flowback/produced water to represent all of the O&G related wastewaters described above.

Treatment requirements associated with selecting/designing a wastewater treatment scheme to be implemented in a nonconventional oil and gas play such as Marcellus are often dictated by a variety of factors, including:

- Regulatory drivers
- E&P/Service company-specific hydrofracturing (frac) methods
- E&P/Service company water quality requirements
- Basin/Area characteristics
- Make-up water chemistry
- Parameter-specific limits for reuse as frac water, including those for:

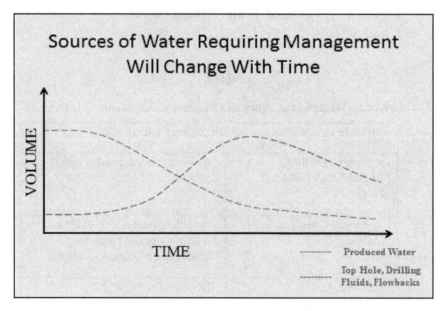

**Figure 2 – Sources of Water Requiring
Management Will Change With Time**

o Total suspended solids (TSS)
o TDS, hardness, alkalinity, chloride

- o Scalants (sulfate, barium, strontium)
- o Bacteria
- o Organics
- o Heavy metals

Typical Characteristics of Marcellus Shale Gas Play Wastewaters

The quality of flowback and produced water varies across nonconventional plays and is also very dependent on operational practices. Most E&P companies active in the Marcellus play are also placing a heavy emphasis on recycle and reuse, which has a direct impact on water quality and associated treatment/disposal objectives/requirements. Table 2 summarizes the typical characteristics of flowback/produced water in Pennsylvania. A particular challenge for the Marcellus play is the relatively high total dissolved solids (TDS) content, which limits available technologies for TDS removal. Another wastewater characteristic-related factor which impacts operation of treatment facilities is the periodic receipt of wastewaters that are relatively high in chemical additives such as friction reducers which can affect the performance of treatment systems and specific unit processes. In addition, as midstream companies develop their assets, the need for management of condensates, which are typically higher in oil and BTEX, will increase.

Table 2 – Typical Characteristics of Raw and Distilled Wastewater Compared to PADEP

Parameter	Typical Profile for Raw Wastewater Generated in the Marcellus Shale Play (mg/L)	Typical Profile for Treated (Distilled) (mg/L)	Dewasting Effluent Limits in Appendix A of PADEP's Revised WMGR123 permit (mg/L)
Aluminum	100 - 200	0.02 - 0.03	0.2
Ammonia	NA	50 - 135	2
Arsenic	NA	20 ug/L	10 ug/L
Barium	100 - 100,000	0.37 - 3.0	2
Benzene	NA	NA	0.12 ug/L
Beryllium	NA	1 ug/L	4 ug/L
Boron	NA	0.02 - 0.03	1.6
Bromide	NA	0.12 - 0.69	0.1
Butoxyethanol	NA	NA	0.7
Cadmium	NA	3 ug/L	0.16 ug/L
Chloride	2,000 - 180,000	8 - 80	25
COD	NA	1,300 - 27,000	15
Chromium	100 - 200	0.01	10
Copper	100 - 200	30 - 40 ug/L	5 ug/L
Ethylene Glycol	NA	16,000 - 50,000 ug/L	13 ug/L
Gross Alpha	NA	3	15 pCi/L
Gross Beta	NA	4	1,000 pCi/L
Iron	100 - 200	0.03 - 0.08	0.3
Lead	100 - 200	20 ug/L	0.13 ug/L
Magnesium	100 - 200	0.04 - 0.2	10
Manganese	100 - 200	NA	0.2
MBAS (Surfactants)	NA	0.05 - 0.06	0.5
Methanol	NA	NA	3.5
Molybdenum	100 - 200	0.01 - 0.03	0.21
Nickel	100 - 200	10 ug/L	30 ug/L
Nitrite-Nitrate Nitrogen	NA	0.4 - 2.3	2
Oil & Grease	NA	10.5 - 16.0	ND
pH	4 - 13	8 - 10	6.5 - 8.5 SU
Radium-226/Radium-228	NA	2.6 - 20	5 pCi/L
Selenium	100 - 200	20 - 30 ug/L	4.6 ug/L
Silver	100 - 200	10 ug/L	1.2 ug/L
Sodium	10,000 - 32,000	3 - 33	25
Strontium	500 - 3,000	0.12 - 2.0	4.2
Sulfate	100 - 1,000	0.17 - 0.82	25
Toluene	NA	NA	0.33
TDS	1,000 - 365,000	1 - 220	500
TSS	>500	1 - 30	45
Uranium	NA	NA	30 ug/L
Zinc	100 - 200	5 ug/L	65 ug/L

NA - Not Analyzed

Typical treatment/management technologies utilized in the Marcellus play can be segregated into the following categories:

- Oil/Water Separation
- Precipitation/Clarification
- Filtration
- Storage
- TDS/Brine Management
- Organic Removal
- Polishing

The number and type of treatment/management technologies employed at any given facility varies based on analysis of the various factors and challenges described previously.

Regulatory Drivers

Regulatory requirements are a very important driver for determining treatment requirements in Pennsylvania. When E&P companies initiated intense shale gas development operations in Pennsylvania in 2008, disposal of flowback and produced waters at publicly owned treatment plants (POTWs) was the common practice. In early 2009, The Pennsylvania Department of Environmental Protection (PADEP) released a document entitled "Permitting Strategy for High Total Dissolved Solids (TDS) Wastewater Discharges". PADEP unilaterally imposed a new method for addressing high TDS discharges. PADEP concluded that total dissolved solids (TDS) is not a primary pollutant, but that it effects the aesthetic qualities of drinking water and it is a potential indicator of chemical constituents in streams and ground water.

In April 2011, amid criticism from environmentalists and growing concern from scientists regarding disposal of O&G related wastewaters, the PADEP asked the state's booming natural gas industry to halt disposing of millions of gallons of flowback/produced water through POTWs that discharge into rivers and streams. The April 2011 announcement was a major change in the state's regulation of gas drilling. It came the same day that an industry group said it now believes drilling wastewater was partly at fault for rising levels of bromide being found in Pittsburgh-area rivers.

The PADEP had concluded that POTWs were ill-equipped to remove pollutants (specifically TDS) from these wastewaters. The PADEP set a May 19, 2011 deadline for drillers to stop bringing the wastewaters to the treatment plants. It did not say how the wastewaters should be disposed of. At the time, the PADEP rules and regulations for disposing of flowback/produced water that is sent off-site for reuse was regulated as a residual waste, requiring permitting under one of three different General Permits applicable to oil and gas operations, as enforced by the PADEP Bureau of Waste Management: WMGR119, WMGR121, and WMGR123.

On March 24, 2012 the Pennsylvania Department of Environmental Protection (DEP) revoked WMGR119 and 121 and revised/reissued General Permit WMGR123, which authorizes the processing and beneficial use of processed liquid wastes generated on oil and gas well sites and associated infrastructure. WMGR123, issued under the authority of the PADEP Bureau of Waste Management, replaced the three existing general permits which previously regulated the recycling and reuse of oil and gas wastewaters.

The new general permit removes some current restrictions on the recycling of oil and gas wastewaters, and also adds some new requirements. For facilities that plan to recycle and reuse relatively dilute waters, the new permit should be helpful. In particular, for wastewaters with low total dissolved solids (TDS) (i.e., less than 500 mg/l) that are in compliance with standards found in Appendix A of the permit (see Table 1), the wastewater will essentially be considered dewasted and E&P companies will not have to manage the waste as a residual waste, and should be able to utilize existing fresh water designs for impoundments and handling of the water.

However, for high TDS wastewaters which do not comply with the Appendix A standards, both the generators and users of the recycled water will potentially have new compliance standards. Until the processed oil and gas liquid wastewater has been transported to a well site and is actually used to develop a well, it must be managed as a residual waste. The requirements to manage the wastewater as a residual waste apply to both the operator generating the waste and the operator reusing the waste. If either the generator of the wastewater, or the party beneficially reusing the wastewater, wishes to store the waste prior to either shipment or reuse, they will need to comply with storage requirements that are generally more stringent than the requirements under the oil and gas regulations.

Eureka Resources, LLC - Case Study

History

Eureka Resources, LLC (Eureka) currently provides centralized treatment, recycling and/or disposal of flowback/produced waters (wastewaters), as well as waste drilling fluids generated as a result of nonconventional natural gas exploration and development activities in the Marcellus play. Eureka placed its first centralized pretreatment facility in operation in November 2008 in Williamsport, PA with a capacity of approximately 4,800 barrels per day (bpd) or 200,000 gallons per day (gpd). The facility was originally permitted by the PADEP Bureau of Waste Management under WMGR119. In 2010, Eureka opened an expanded centralized treatment facility which is currently capable of treating/recycling up to 10,000 bpd (420,000 gallons) of flowback/produced wastewater daily. The facility permit was converted to a WMGR123 permit at this same time, as authorized by the PADEP Bureau of Waste Management.

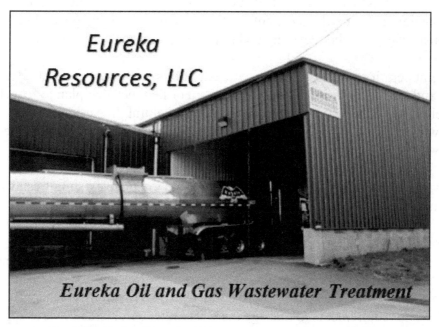

**Figure 3 – Eureka Oil and Gas Wastewater
Treatment**

Eureka has the most comprehensive centralized wastewater treatment capabilities in the
Marcellus play. Eureka's customers currently have the choice of requesting various levels
of treatment with the goal of maximizing recycle of:

- Pretreated Water - water treated to reduce the suspended solids and heavy
  metals content and readily-settleable dissolved constituents.
- Distilled Water – water treated to remove more soluble dissolved solids,
- Concentrated Brine – water very high in TDS which can be reused as a
  drilling
  fluid additive.

Future Plans

Eureka is in the process of planning, permitting, designing, and constructing facilities to provide the following additional wastewater treatment/disposal services in support of nonconventional E&P activities below and on Figure 4:

- Concentrated Brine Management Options:
    - o Further centralized treatment using mechanical vapor recompression (MVR) crystallizers, with beneficial reuse of a dewatered sodium chloride and purge stream byproducts
    - o Deep well injection in Ohio and/or West Virginia

Figure 4 – Eureka's Business Plan

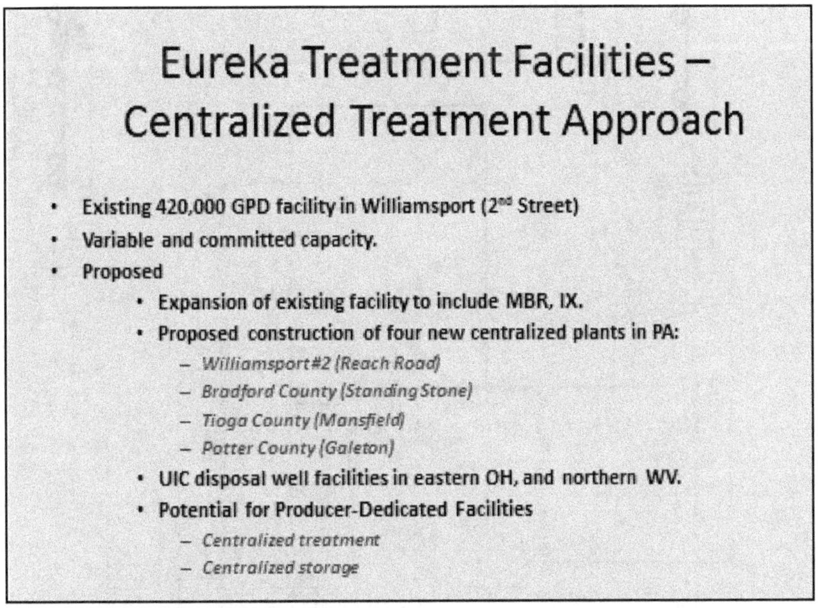

Eureka Treatment Facilities –
Centralized Treatment Approach

- Existing 420,000 GPD facility in Williamsport (2ⁿᵈ Street)
- Variable and committed capacity.
- Proposed
    - Expansion of existing facility to include MBR, IX.
    - Proposed construction of four new centralized plants in PA:
        - Williamsport #2 (Reach Road)
        - Bradford County (Standing Stone)
        - Tioga County (Mansfield)
        - Potter County (Galeton)
    - UIC disposal well facilities in eastern OH, and northern WV.
    - Potential for Producer-Dedicated Facilities
        - Centralized treatment
        - Centralized storage

- Tertiary treatment of distilled water using membrane biological reactors (MBR) followed by reverse osmosis (RO) to generate dewasted water for unrestricted reuse [Patent Pending].
- Establishment of three more centralized treatment facilities in Pennsylvania, as well as possible additional centralized treatment facilities in Ohio and West Virginia.
- Dedicated producer centralized or satellite treatment and/or storage facilities.

A process flow diagram for Eureka's concept for a comprehensive centralized treatment facility is illustrated on Figure 5.

145

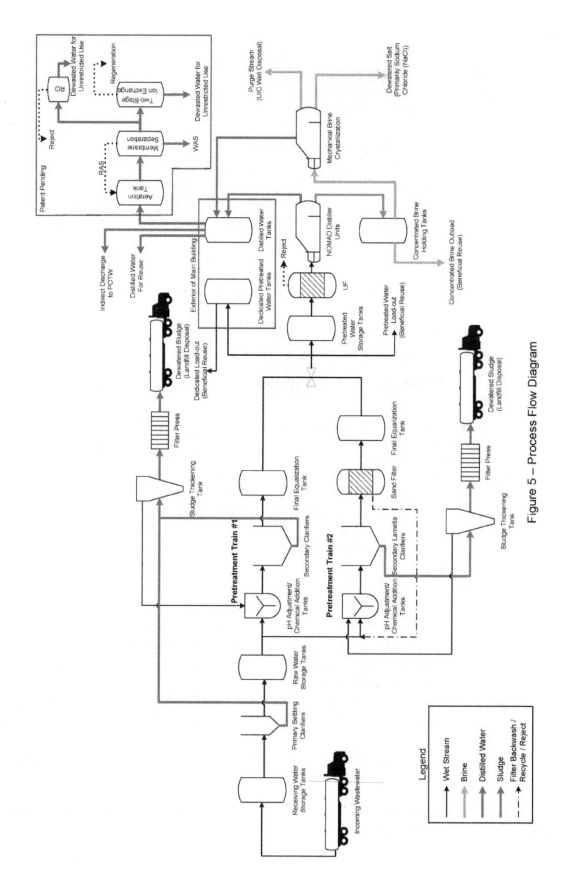

Figure 5 – Process Flow Diagram

146

Process flow descriptions for the unit processes shown on the process flow diagram are as follows:

- **Initial Pretreatment:**
  - Receiving Wastewater – Before any raw, untreated wastewater is unloaded into partially below grade concrete receiving water storage tanks, it is evaluated by one of the Eureka operators. The water is tested for total dissolved solids (TDS), barium, sulfate, pH, settleable solids, and also visually inspected for oil and other visual indicators. If it passes the visual inspection, the truck is unloaded into one of four concrete receiving water storage tanks at the direction of the plant operator.
  - Primary Settling Clarifiers/Oil Trap – Situated above the concrete receiving water storage tanks are three steel clarifiers, which are used to collect and remove rapidly-settling solids and trap oil. Prior to pumping water into these tanks, the large mixers in the concrete receiving water tanks are turned on to fluidize the contents. The slurry is then pumped into one of the overhead clarifiers. The heavy solids settle in these clarifiers. The liquid flows over a weir into one of the partially below grade concrete receiving water storage tanks. This tank is equipped to serve as a raw water transfer tank, including a pump to transfer the water into one of the indoor storage tanks. The weirs on the primary clarifiers are set up such that any free oil in the clarifier is collected on the surface of the tank. Once a noticeable layer of oil collects on the surface of one of these clarifiers, it is drained into an oil-water separator and thus prevented from entering the downstream unit processes.
  - Indoor Raw Water Storage Tanks – Each of eight indoor above grade steel raw water storage tanks are equipped with a designated fill line from an overhead manifold that allows operators to select which tank to fill from the pump in the raw water transfer tank. When in operation, these raw water tanks can also be isolated from the process via isolation valves along a common header that supplies a process feed pump. These two features enable operators to selectively fill and draw from a given tank or set of tanks in order to manage variable influent water quality. The process feed pumps deliver water from the specified raw water tank into pH adjustment/chemical addition tank No. 1, where chemical treatment begins.
    - Over time, some solids settle/accumulate in the raw water storage tanks. There are two progressing cavity pumps connected to the header at the bottom of the raw water storage tanks that are used to transfer solids from the tanks directly into the sludge holding tanks.
    - There is a third header that is used to put water into each tank individually from the final equalization tank in the process. If there is a process upset, this allows operators to recycle water from the

end of the process back to the raw water tanks in order to manage pre-treated effluent that does not meet specifications.

- **Pretreatment Train #1:**
  - pH Adjustment/Chemical Addition Tanks – Prior to the secondary clarifier, there are three 8,000-gallon pH adjustment/chemical addition tanks labeled pH tanks 1, 2 and 3.
  - Acid (hydrochloric) or alkali (caustic soda or lime) is added to each of these tanks in order to lower or raise the pH when necessary to achieve proper metals removals/settling. Sodium sulfate can also be added to the first pH adjustment/chemical addition tank in order to promote precipitation of barium. The second pH adjustment/chemical addition tank is used for pH polishing only. In the third pH adjustment/chemical addition tank, a coagulant is added in order to aid in the removal of suspended particles. Polymer is also added to the 250-gallon flocculation tank downstream of the three pH adjustment tanks to further aid settling.
  - Secondary Clarifiers/Filtration – Following pH adjustment and chemical addition, the process water flows into one large circular clarifier. The solids fall to the bottom of the clarifier before being pumped into the sludge holding tanks. The resultant clarified pretreated water rises to the top of the clarifier and is conveyed to the final equalization tank.
  - Sludge Thickening Tanks – The solids from the bottom of the clarifier drain into a sump and are then pumped into the first of three sludge thickening tanks. In this tank, the solids compact toward the bottom of the tank before being pumped into the second sludge thickening tank. The supernatant water that rises to the top of the thickening tank is decanted and directed back to pH tank 2. In the second and third sludge thickening tanks, the solids thicken further and are then pumped from the bottom of the tanks to the filter press. The supernatant waters from these tanks also decant back to pH tank 2.
  - Rotary Press – The thickened sludge is pumped using a progressive cavity pump to a dewatering rotary press. The solids in the sludge are dewatered by the rotary press and, through the course of the cycle, compacted into a solid filter cake. The rotary press allows for continual steady stream dewatering, unlike a traditional filter press which is a batch process, and requires typically only a once daily cleaning cycle. The produced cake drops into the sludge roll-off below. Once filled, the roll-off is hauled away for disposal and a clean empty one is put in its place. The filtrate from the filter press is directed to the pH tank 2.
  - Final Equalization Tank (Pretreated Water Transfer Tank) – The last unit in the pretreatment process is the final equalization tank. The final equalization tank serves as a pretreated wastewater transfer tank where pretreated wastewater will be pumped to one of two destinations:

148

- o   one of ten pretreated water storage tanks
- o   pretreated water load-out conduit at the main, covered bulk transfer area.

- **Pretreatment Train #2:**

  - pH Adjustment/Chemical Addition Tanks – Prior to the secondary clarifier, there are three 6,000-gallon pH adjustment/chemical addition tanks labeled pH tanks 4, 5 and 6. Acid (hydrochloric) or alkali (caustic soda or lime) can be added to each of these tanks in order to lower or raise the pH when necessary to achieve proper metals removals/settling. Sodium sulfate can also be added to the first pH adjustment/chemical addition tank in order to promote precipitation of barium. The second pH adjustment/chemical addition tank is used for pH polishing only. In the third pH adjustment/chemical addition tank, a coagulant is added in order to aid in the removal of suspended particles.

  - Secondary Lamella Clarifiers – Following pH adjustment and chemical addition, the process water flows into one of two above grade steel inclined plate lamella clarifiers. The solids fall to the bottom of the clarifier before being pumped into the sludge holding tanks. The resultant clarified pretreated water rises to the top of the clarifier where it flows to the sand filter or directly to the final equalization tank.

  - Sludge Thickening Tanks – The solids from the bottom of the clarifiers drain into a sump and are then pumped into the first of three sludge thickening tanks. In this tank, the solids compact toward the bottom of the tank before being pumped into the second sludge thickening tank. The supernatant water that rises to the top of the thickening tank is decanted by gravity back to pH tank 5. In the second and third sludge thickening tanks, the solids thicken further and are then be pumped from the bottom of the tanks to the filter press. The supernatant water from these tanks also decant back to pH tank 5.

  - Filter Press – The thickened sludge is pumped using a diaphragm pump to the filter press. The solids in the sludge are dewatered by the filter press and through the course of the cycle are compacted into a solid filter cake. At the end of the press cycle, the press is opened and the cake drops into the sludge roll-off below. Once filled, the roll-off is hauled away for disposal and a clean empty one is put in its place. The filtrate from the filter press is directed to the pH tank 5.

  - Sand Filter – The supernatant from the lamella clarifiers can be directed to a continuously back flushing sand filter, or directed to the final equalization tank if it already meets discharge criteria. The sand filter provides water polishing by catching any large particles that carry over from the clarifiers. The filtrate passes on to the final equalization tank. The sand filter also generates a constant reject stream containing the solids removed by the filter. This reject stream is pumped back to pH tank 5 to re-enter the process.

149

- Final Equalization Tank (Pretreated Water Transfer Tank) – The last unit in the pretreatment process will be the final equalization tank. The final equalization tank will serve as a pretreated wastewater transfer tank where pretreated wastewater will be pumped to one of two destinations:
  - one of ten pretreated water storage tanks
  - pretreated water load-out conduit at the main.

Figure 6 summarizes the process performance data for Eureka's existing pretreatment train.

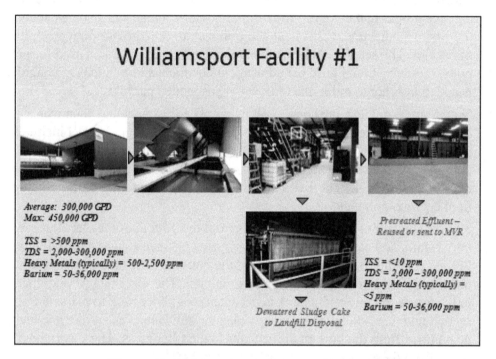

Figure 6 – Pretreatment Process Performance Data

- **Secondary Treatment:**
  - Pretreated Water Holding Tanks – Each of the above grade steel pretreated water storage tanks is equipped with a designated fill line from an overhead manifold that allows operators to select which tank to fill from the pump at the pretreated water transfer tank. When in operation, these tanks can also be isolated from the process along a common header that supplies a process feed pump. These two features enable operators to selectively fill and draw from a given tank or set of tanks. The process feed pumps deliver water from the pretreated water tanks to the mechanical vapor recompression (MVR) distiller units. .
  - **MVR Distiller Units** – If advanced treatment is needed, pretreated water is pumped from the pretreated water holding tanks to the MVR distiller units. The feed water passes through a feed filter where any residual solids are removed. The

feed water flow is then split into two streams and directed to the feed/distillate preheat exchanger and feed/concentrate preheat exchanger with the feed control valves. The feed water heats up to near boiling as it recovers sensible heat from the hot distilled water and hot concentrate leaving the MVR distiller. After the preheat exchangers, the two feed water streams combine to a single stream and enter the de-aerator, where non-condensable gases are stripped from the feed water and vented to atmosphere.

The feed water then drops to the bottom of the de-aerator column, where it combines with a large volume of hot re-circulating concentrate. The feed water blends with the re-circulating concentrate and then pulled out of the separator vessel through a nozzle beneath the de-aerator by recirculation pumps. The hot re-circulating concentrate is driven from the recirculation pumps up through the evaporator exchanger where approximately 5% of the liquid is vaporized into steam on the outside of the exchanger cassettes. The mixture of steam and hot concentrate exits the top of the evaporator exchanger and flows into the separator vessel through the inlet nozzle. Inside the separator vessel, the hot concentrate separates from the steam by gravity. A vane pack type mist eliminator inside the vessel removes small particles of liquid from the steam.

The steam is then drawn from the separator vessel through the steam discharge nozzle by the compressor. The compressor boosts the pressure and temperature of the steam by approximately 6 to 8 psi and 18 to 23° F. The hot, high-pressure steam is driven from the discharge of the compressor to the evaporator exchanger. The steam enters the evaporator exchanger through the top. As the steam moves down through the inside of the exchanger cassettes, it condenses into distilled water. The latent heat energy given up by this compressed steam as it condenses transfers to the circulating brine vaporizing an equivalent amount of brine. The amount of energy required to vaporize the brine is equal to the energy needed to compress the steam.

The hot distilled water is collected in the distillate receiver. The discharge pressure of the compressor and steam is determined by the temperature approach required to maintain boiling in the exchanger. Operating pressure in the separator vessel is maintained by regulating the excess steam from the distillate receiver to hold separator pressure between 0.75 and 1.5 psi.

Hot distilled water is pumped from the distillate receiver with the distillate pump through the feed/distillate preheat exchanger. Inside the feed/distillate preheat exchanger, the distilled water cools down, passing sensible heat to the incoming feed. The distillate water then passes through the compressor oil cooler, where the cool distilled water is used to cool the compressor oil. A backpressure valve maintains pressure on the distilled water to verify that the distillate pump is operating properly. The distilled water flows out of the distiller to the treated effluent/distillate transfer tank, from which Eureka has the option to either pump it to the treated effluent/distillate storage tanks, or discharge to the Williamsport

Sewer Authority's collection system in accordance with an indirect discharge permit.

A start-up boiler is used to heat up the distiller unit prior to start-up. The boiler is fed distilled water from the distillate receiver via the distillate pump. The boiler boils the distilled water into steam. This steam directed to the steam piping at the inlet of the evaporator exchanger condenses inside the evaporator exchanger, passing its latent heat of condensation to the re-circulating concentrate, causing the system to warm up. The start-up boiler is shut down after the system is warmed up.

- Concentrated Brine Holding Tanks - To maintain a constant specific gravity of re-circulating concentrate, a concentrate (or blow down) stream is drawn from the separator vessel with the concentrate pump. The hot concentrate flows to the feed/concentrate preheat exchanger, where the concentrate cools down, passing its sensible heat to the incoming feed. The cool concentrated brine then is pumped to one of the four concentrated brine holding tanks.

Figure 7 summarizes the process performance data for Eureka's existing secondary treatment train. Table 2 presents a detailed summary of distilled water quality.

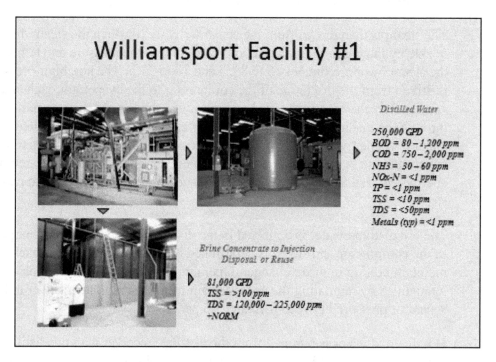

Figure 7 – Secondary Process Performance Data

152

- **Additional Storage/Future Treatment:**

  - **Pretreated Effluent/Distillate Storage Tanks** –Two, double-walled 1,500,000-gallon pretreated effluent storage tanks and three, double-walled 145,000-gallon distillate storage tanks are utilized. Pretreated effluent generated by the facility is pumped to the pretreated effluent storage tanks to await bulk transfer/sale offsite to E&P service companies for beneficial reuse for various Marcellus shale gas drilling operations. These tanks can be dedicated to one contracted customer for these purposes. Distilled effluent produced by the MVR distiller units is pumped to the distilled water storage tanks to await bulk transfer/sale offsite to E&P services companies for beneficial reuse for various Marcellus shale gas drilling operations.

  - **Mechanical brine crystallization** – Eureka's current WMGR123 residual waste management general permit issued by the PADEP includes approval to add an electric-driven mechanical vapor recompression (MVR) crystallizer unit at the treatment facility. The unit would be used primarily to further process/dewater the brine concentrate to be produced by the MVR distiller units. Crystallizer performance targets are presented on Figure 8.

    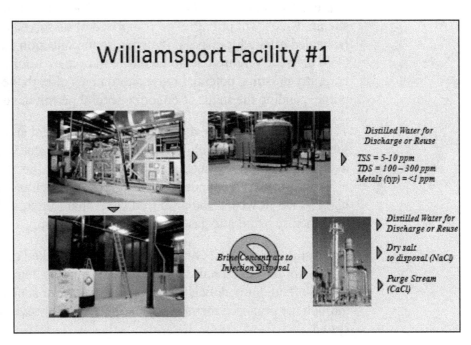

    Figure 8 – Crystallizer Performance Targets

    The crystallizer unit will be comprised primarily of a mechanical vapor recompression forced circulation-type evaporator system. The system will be fully integrated and automated using mechanical vapor recompression to concentrate high-solubility constituents of the pretreated wastewater feed produced from the physical-chemical pretreatment process proposed as part of Phase I.

The pretreated wastewater fed to the unit (feed liquor) will first be passed through a pair of heat exchangers to gather heat from the heated distilled water and liquid brine purge discharged from the process, prior to be being fed to the forced circulation evaporator body. Inside the evaporator body, the feed will be heated in

153

the heating chest portion of the evaporator. The heated feed will be directed to the evaporator body where boiling will occur and sodium chloride salt crystals will form, which will fall down through the salt elutriation leg via gravity to form a salt bed. To maintain a favorable set of conditions for the re-circulating liquid within the evaporator, a liquid brine purge stream (mother liquor) will be drawn from the evaporator vessel at this point in the process. This allows for greater control of the sodium chloride salt crystal formation, resulting in crystals of consistent size and high purity. The salt slurry from the elutriation leg will be directed to a pusher-type centrifuge, which will dry the salt to approximately 97 to 98% solids.

The vapor leaving the evaporator body is passed through a demister prior to being fed into three blowers and recompressed as heating vapor in the heating chest of the evaporator (i.e., used to preheat incoming feed). The cooled condensed vapor (distilled water) is then collected.

Following centrifugal dewatering, the dry salt cake will either be transported to a solid waste landfill for disposal, or stored onsite in the proposed dry salt storage building to prepare it for beneficial reuse. The concentrated liquid brine purge stream, which will primarily be comprised of dissolved calcium chloride, will either be hauled offsite for disposal in underground injection disposal wells (likely in Ohio or West Virginia) or stored onsite to prepare it for beneficial reuse. Eureka is currently pursuing potential reuse alternatives of both the dry salt and liquid brine purge, pending the status of ongoing general permit development by the PADEP.

The crystallizer will also allow Eureka to implement an alternative treatment scheme for higher-TDS flowback and produced water which would involve pre-treatment followed by treatment in the crystallizer (i.e., bypass of the distiller units). Use of the crystallizer under this treatment approach will allow for treatment of higher-TDS flowback and produced water that Eureka projects will be brought to their facility by oil and gas developers in the region.

The distilled water will either be immediately trucked offsite by licensed/permitted haulers for beneficial reuse by gas well developers, temporarily stored onsite to be hauled offsite for beneficial reuse at a later date, or subjected to further treatment through the proposed membrane biological reactor/reverse osmosis (MBR/RO) system.

- **Membrane Biological Reactor (MBR)/Reverse Osmosis (RO) treatment** [Patent Pending] – The MBR/RO system will allow Eureka to further treat the distilled water (see Table 2 for distilled water quality) produced by the crystallizer using a biological treatment processes similar to those that are typically used at most municipal wastewater treatment plants. Following treatment via these processes, the distilled water will meet the dewasting treatment standards set forth in Appendix A of the revised WMGR123 permit recently issued by the PADEP (i.e., it will no longer be regarded as a residual waste). This will allow for storage of the effluent in an impoundment or other facility under the control of an oil and gas

154

producer prior to reuse as frac water, without required compliance with storage and transportation requirements in 25 PA Code Chapter 299. The distilled, dewasted effluent will either be immediately trucked offsite by licensed/permitted haulers for unrestricted beneficial reuse by gas well developers, or temporarily stored onsite to be hauled offsite for unrestricted beneficial reuse at a later date.

The MBR system will be designed to provide biological removal of chemical oxygen demand (COD) and ammonia in the distilled water feed. The MBR system will consist of feed pumps, a mixed anoxic tank, two aeration tanks, the membrane separation system, sludge recycle pumps, an automatic sludge wasting system, and final effluent pumps.

The pending potent covers use of both RO and Ion Exchange as polishing steps. For this paper we have focused on the RO approach. The RO system will be used as a polishing step downstream of the MBR system to remove residual nitrate and sodium that may be in the MBR effluent.

In the RO process, water under pressure will be forced across a membrane element, with a portion of the feed permeating (by diffusion) the membrane ("permeate"), and the balance of the feed water sweeping along the membrane surface and exiting without passing through the membrane ("reject"). During the diffusion process, the membrane will freely pass the water molecules but will reject most of the dissolved salts and metal ions, the small particles as well as organic compounds and some bacteria.

The reverse osmosis process will typically convert, or "recover", a certain percentage, between 50 to 80%, of the initial incoming feed water into "RO permeate". The remaining by-product, or the "RO concentrate", will contain/concentrate the salts rejected by the membrane.

Reuse Strategy

Eureka's strategy focuses on providing a centralized treatment facility for flowback and produced wastewaters that allows maximum flexibility for applying the level of treatment necessary to promote reuse of the treated water (see Figure 9). Using this approach, Eureka is able to provide the following for potential reuse by E&P companies:

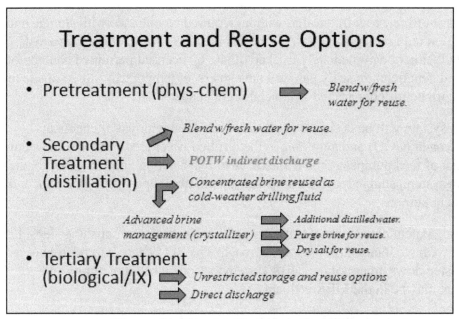

Figure 9 – Eureka Treatment and Reuse Options

- Pretreated Water
- Distilled Water
- Dewasted Water
- Concentrated Brine Purge
- Dry Salt (< 3% moisture)

Conclusions/Takeaways

Actual experience treating flowback and produced wastewaters generated in Pennsylvania as a result of the Marcellus Shale Gas play has indicated the following:

- In the highly competitive US shale gas market, it is important for oil and gas (O&G) exploration and production (E&P) companies to find efficient solutions to one of the high cost aspects of production; treatment of flowback and produced waters.
- E&P companies face many challenges when developing and implementing water management/treatment/disposal strategies including variable water quality, need to establish treatment objectives, need to evaluate treatment technologies/schemes and need to adjust the approach to address operational issues and regulatory requirements. It is important to recognize that treatment requirements vary spatially and temporally.
- Wastewater from nonconventional O&G development is comprised of a variety of constituents that are difficult and costly to treat; direct treatment at

156

most POTW's is not an option. A particular challenge for the Marcellus play is the relatively high total dissolved solids (TDS) content which limits available technologies for TDS removal.

- The Pennsylvania Department of Environmental Protection (PADEP) recently revoked General Permits WMGR119 and 121 and revised/reissued General Permit WMGR123 to govern the processing and beneficial use of processed liquid wastewaters generated on oil and gas well sites and associated infrastructure. The revised General Permit WMGR123 established treatment requirements for dewasting these wastewaters.

- The O&G industry as a whole is developing best available technology (BAT) to meet treatment requirements. Many BAT strategies are basin specific.

- Eureka has over four years of successful experience treating complex unconventional Marcellus play flowback and produced waters generated in Pennsylvania, by providing the following to their customers:
  - Wide range of treatment levels available; allows for flexible wastewater treatment strategy.
  - Focus on selecting the level of treatment to optimize reuse options and reducing impact on hydrologic cycle.
  - Forward-thinking strategic position to stay at forefront of emerging regulatory requirements; close dialogue with state regulators regarding the development of regulations and standards.

References

- General Permit WMGR123, Pennsylvania Department of Environmental Protection, March 24, 2012
- "Permitting Strategy for High Total Dissolved Solids (TDS) Wastewater Discharges", Pennsylvania Department of Environmental Protection
- "Modern Shale Gas Development in the United States: A Primer", U.S. Department of Energy, April 2009

LINER CONSIDERATIONS AT UNCONVENTIONAL DRILLING SITE IMPOUNDMENTS

Veronica E. Foster, P.E., Golder Associates Inc., Mount Laurel, New Jersey

One of the many considerations when advancing natural gas wells for unconventional drilling is protection of the environment in the event of a spill or leak. While regulations governing containment on unconventional drilling pads vary by state, the goals and objectives of the natural gas well development industry remain the same: minimizing risk in a cost effective manner. To minimize the risk of a spill or leak, engineering controls are the first measures to be considered. These engineering controls may be active or passive and consist of any combination of the following controls:

- Containment berms around all or part of the well pad;
- Low-permeability barriers on all or part of the well pad;
- Secondary containment on tanks (permanent or temporary);
- Liquids management controls;
- Spill kits; and,
- Changes in the fluids used for drilling operations.

Any combination of these engineering controls may be used, based upon the owner's preferences and the governing state regulations. However, when selecting the engineering controls, a number of factors require evaluation, including but not limited to, the following:

- Health and safety of workers on the pad
    - Will special PPE be required?
    - Is there increased risk of slips, trips, falls due to the control?
    - Is the impoundment or pit sufficient deep that access should be restricted?
    - Are weather conditions like rain, snow, ice, electrical storms a consideration?
- Risk of leak in that particular location
    - Common - hydraulic fluid from a mobile engine
    - Typical – fluid transfer between containers
    - Unlikely – tank rupture
- Size of the possible leak

- o Small (<1 gallon) – drops of hydraulic fluid; tank leakage
  - o Medium (<55 gallons) – spillage during fluid transfers, hose or drum rupture
  - o Large (>55 gallons) – tote or tank rupture
- Downstream receptors and their classification
  - o Stream – high quality or exceptional value
  - o Wetland – exceptional value, low quality
  - o Protected Water Supplies – ponds for watering cattle (if in farmland)
- What is the nature of the soils comprising the well pad? The type of the natural soils may help contain a spill or may help mobilize it
  - o Clayey
  - o Sandy
  - o Stone
- What is the climate in which the well pad is located and what season?
  - o Arid (e.g., West Texas, Montana)
  - o Sub-tropical humid (e.g., Gulf states)
  - o Humid Continental (e.g., Wisconsin)

Once these factors are assessed, the locations and types of engineering controls are determined. The materials that may be used in the key elements of the engineering controls, the low permeability barrier, are numerous. The materials can be natural or man-made. Clay is often the "natural material" choice for low permeability barriers. However, there are many parts of the US (and the world) where appropriately clayey soils are not locally available, and the cost to import them to the work area is cost-prohibitive. In those instances, there is a wide range of geosynthetic products which may be used. These products include:

- High Density Polyethylene (HDPE)
- Linear Low Density Polyethylene (LLDPE)
- Polypropylene (rubber-based) (fPP)
- Poly Vinyl Chloride (PVC) (North American manufactured material)
- Ethylene Propylene Diene Monomer (EPDM)
- Geosynthetic Clay Liner (GCLs)

Photograph by Veronica Foster

Each of these products functions well as a low permeability barrier, but each one offers different positive characteristics. Many of these materials are sold in rolls, ranging in width of 15 feet to 22 feet and ranging in lengths of 300 feet to 520 feet, depending upon the material. As such, the material is deployed by rolling it out in panels, and cutting the

160

material to fit the geometry of the installation location. Adjacent panels are welded together using any of three different types of welding equipment.

GCLs differ from the other geomembrane materials listed above as the GCLK is a combination of a geosynthetic fabric combined with naturally occurring bentonite (or powdered clay) and the geomembranes are engineered plastic or rubber-based products. The GCL panels may simply be overlapped, with additional bentonite added along the panel edges. Since GCLs act as a low permeability barrier by swelling as a result of absorption of liquid, GCLs serve better as secondary liners situated beneath one of the other geomembrane barrier liners. Therefore, GCLs are not discussed herein any further.

When selecting a geomembrane or geosynthetic product, the following factors need to be considered:

- To what chemicals will the material be exposed? (i.e. chemical compatibility)
- For how long is this material expected to be in service? (i.e. 6 months, 6 years, 16 years?)
- What is the physical configuration of the area where the material will be installed?
- What stresses/loads will be imposed upon the material?
- What are the installation challenges?
- What do these material cost?
- What is the time needed to obtain the materials?
- Are there specialty installation protocols for the materials?
- Who is qualified to install the geosynthetic materials?
- Are there specialty manufacturers who cater to the oil & gas industry?

Chemical compatibility is a question that will have to be addressed by the driller or oil & gas company, as many of the drilling chemicals are proprietary.

Duration of application depends upon the number of wells being advanced, the shape and size of the area being lined with the geosynthetic material, and the environmental conditions to which the liner are exposed. Assuming that all liner applications on a well pad are exposed, the longevity of the various liner materials due to sun light (ultraviolet exposure), as reported by the Geosynthetics Institute (GSI) based upon laboratory testing, is as follows:

Material Type	Predicted Exposed Lifetime in Arid Climate
HDPE	50+ years
LLDPE	33 years
fPP	33 years
EPDM	30 years
PVC	7 years

Based upon the above table, PVC as manufactured in North America is not recommended for long-term exposed applications. The material degrades due to ultraviolet exposure relatively quickly when compared to other materials.

The geometry of the area being lined with the material has a significant impact of the material selection. If the material is being placed on a flat area, any of the material outlined above would perform adequately. For locations with steep slopes to be lined, the material may need to be reinforced to withstand the stresses being imposed by the installation and operation loads. In the instance of open-air water impoundments or drill pits, textured material is recommended for several reasons. The texturing will improve the stability of the material and its interface with the subgrade materials/soils, and will improve the traction on the walking surface for the installer and ensuing operations personnel.

When looking at impoundments or drill pits, the load, or head (i.e., the thickness of water and solids), atop the liner must be assessed. The amount of liquid atop the liner will dictate the minimum material thickness that needs to be installed. For example, your state's regulations may specify a minimum thickness of 30 mils for a geomembrane. However, with 15 feet of liquid atop the material, engineering analysis may stipulate that a 40-mil thickness is required. The opposite may also occur, wherein the engineering analysis indicates that only a 20-mil thickness is sufficient. In this example, a minimum 30-mil thickness would be used to comply with the state regulations.

However, the loads imposed atop the liner may not be the critical condition to be evaluated. The discharge from the drilling operation into the impoundment or pit, may have such force, and may have enough particulate matter, that could create the equivalent of a "sand blaster." The stresses on the liner system from this pocket of increased abrasion could be addressed separately from the rest of the area being lined and handled with special protections.

Another factor to consider is the challenge of installing any of these materials. First, most geosynthetic manufacturers will require that "certified" installers must be used. Manufacturers often include this stipulation to prevent persons without demonstrated experience with these products from performing the installation. Often, manufacturers will accept installers which are certified by other manufacturers or by the International Association of Geosynthetics Installers (IAGI).

Second, the installation process may impose more short-term stresses on the material than the long-term operational use. For example, installing a thinner material may take more time because the material is less forgiving than a more robust (i.e. thicker) material. By opting to increase the thickness of the material, the installer may be able to significantly decrease the installation time.

The choice to increase material thickness to expedite installation requires an understanding of the balance between material and installation costs. Doubling the

163

material thickness does not double the cost, but the increase in material cost, may readily be offset by the reduced installation cost. Also, if the well pad construction project schedule is fast-tracked, as they so often are, there is very little cushion for schedule slippage. As such, the increased installation speed may be worth a slight increase in material costs.

For example, the table below shows typical material property values for several textured HDPE geomembrane material thicknesses:

Property*	30 mil	40 mil	% of 30 mil	60 mil	% of 30 mil
Average thickness	30 mil	40 mil	127%	60 mil	200%
Minimum thickness	27 mil	36 mil	126%	54 mil	200%
Tear resistance	21 lb	28 lb	133%	42 lb	200%
Puncture resistance	45 lb	60 lb	133%	90 lb	200%

* Based upon published data from GSE, Houston, Texas.

Compared to the use of a 30-mil thick HDPE geomembrane, the use of a 40-mil thick material provides only marginally greater geomembrane strength and tear and puncture resistance, while the use of a 60-mil thick material provides double the geomembrane strength and tear and puncture resistance. These material properties may be more critical during the installation of the material, than during long-term operations.

Material cost, as always, is dependent upon availability, which is directly linked to demand for the product and the amount of material manufactured. Materials with limited production typically carry longer lead times. While 30-mil and 40-mil material thicknesses may use less resin, there may be fewer orders for the material, such that the manufacturers may offer limited production of the material and only if they have orders in-hand. As a result, this material may, in fact, cost more than another thicker material that is produced daily, and possibly on more than one production line. It is critical to work with an engineer who understands the market indicators, and can work with industry to negotiate the balance between material costs and installation costs for the proposed long-term and short-term applications.

Selecting and purchasing the low permeability barrier or containment liner that works for your application is only half of the equation of minimizing risk. The second half is the installation and documentation of the installation.

For water impoundment and drill pits a Construction Quality Assurance (CQA) Plan should be in place. The CQA Plan outlines the geosynthetic liner installation with the following steps:

- Receipt and inventory of the material;
- Deployment;
- Welding adjacent panels;
- Testing the adequacy of the welds; and,
- Anchoring the panels at the crest of the slopes.

The installation needs to occur under conditions when the ambient temperature is more than 30 degrees F, winds (including gusts) are less than 20 miles per hour (mph), and it is neither raining nor snowing. Installation is restricted during any level of wind causing blowing dust, because it can compromise the quality of welds. If installation must advance under any of these conditions, portable tents and heaters can be mobilized to shield the deployment and welding activities from unfavorable weather conditions.

Photograph by Veronica Foster

The purpose of installing containment for impoundments and pits is to help manage (environmental) risk. What happens when someone files a claim of environmental harm? How do you prove that all of the installation protocols were followed? How do you prove the strength of the welds? How do you prove that the liner did not have any leaks when installed? How do you prove the installation was performed in accordance with the permit-approved design plans, if required in your jurisdiction, and technical specifications?

The answer to the foregoing questions is: *CQA documentation*. Most certifications, however, require the documentation to be developed by an independent observer, and signed and sealed by a licensed professional engineer who is technically and duly qualified, and will attest in court to the installation protocols followed in the field. Typical CQA documentation of a liner installation in a drill pit or water impoundment would include the following components:

- Written narrative describing, in detail, the field protocols followed;
- Independent laboratory test results that indicate that the material meets the project specifications;
- Observations of the material deployment, noting the roll identification of material from which each deployed panel was derived;
- Proof of installer's certification by the manufacturer or the IAGI;
- Proof of the welding apparatus operators qualifications;
- Demonstration, before beginning welding adjacent panel together, that the welding equipment and operator are functioning properly on each day of welding activities;
- Independent observation of welding activities;
- Independent observation of weld testing activities;
- Independent observation of weld repair activities;
- Independent leak testing of the finished lined area; and,
- Certification by a professional engineer, licensed in the state where the work occurs, that these efforts were performed in compliance with the project specifications.

The final test of the liner installation is the independent leak testing. Leak testing is performed with an electrical charge introduced around the perimeter. The surface of the liner is moistened with water, and a "detection wand" is run across every inch of the liner. If the wand detects an electrical charge (i.e., the circuit closes), there is a leak. While the testing of the welds is performed during installation, this testing does not detect defects in the liner material between the welds. Any leaks detected by this method are repaired, and the area re-tested to assure the repair was properly completed.

Photograph by Veronica Foster

By compiling this construction documentation in the form of a clear and concise report upon the completion of liner installation, the "independent documentation" is readily available if any environmental claims are filed.

The inclusion of a qualified engineer, knowledgeable about unconventional well development practices and geosynthetics design and installation, is paramount to the process of helping the oil and gas industry minimize risk, while maximizing profits.

CPSIA information can be obtained
at www.ICGtesting.com
Printed in the USA
LVHW060536030720
659558LV00006B/203